SCIENCE SKETCHES

SCIENCE SKETCHES

The Universe from Different Angles

Sidney Perkowitz

JENNY STANFORD
PUBLISHING

Published by

Jenny Stanford Publishing Pte. Ltd.
Level 34, Centennial Tower
3 Temasek Avenue
Singapore 039190

Email: editorial@jennystanford.com
Web: www.jennystanford.com

British Library Cataloguing-in-Publication Data
A catalogue record for this book is available from the British Library.

Science Sketches: The Universe from Different Angles

ISBN 978-981-4877-94-7 (Hardcover)
ISBN 978-1-003-27496-4 (eBook)

To my beloved wife Sandy and my wonderful family, Mike, Erica, and Nora—I'm grateful that you're here.

Contents

---

*Article written for this book.

Preface: The Universe from Different Angles

This book is the second collection of my published articles and essays. It follows the 50 articles in *Real Scientists Don't Wear Ties* (Jenny Stanford Publishing, 2019) with 52 more pieces. They appear in formats from quick reads to deep dives, and cover topics from quantum gravity to science in the media, from neuroscience to technology in society. This new collection contains mostly items from the last decade right up to this year, but you'll also find older pieces from as far back as the 1990s (for some of these articles where an update to a fact or event would be helpful or interesting, I have provided that in a brief note).

I selected the entries with an eye toward variety in chronology as well as coverage. The topics range broadly in time from the historical and enduring to the latest science and technology and their current effects. Except for two pieces I wrote specifically for this book (see Contents), what I present here comes from varied print and online sources with different goals and styles; all the articles, however, have been written for general readers. Their lengths cover a range, too, from 600 words to over 3000, and these are distributed throughout the book. One difference from *Real Scientists Don't Wear Ties* is that I sometimes place pieces about science in the media alongside articles that directly treat science and technology, as a way to broaden and enliven the discussion.

In *Real Scientists* I related how I became both scientist and writer with a career in physics research and teaching, followed by a second career in writing that draws on my scientific training. I won't repeat that history, except to say again that I have been fortunate in combining two personal passions into one satisfying whole. This became especially clear during the time when the COVID-19 pandemic brought our society to the point of lockdown. For myself, I'm thankful that I have been able to continue writing under these conditions. This kept my mind and attention engaged and contributed to my mental health during these difficult times.

I had structured the 50 pieces in *Real Scientists* by organizing them into three main categories: science, meaning discussions about pure science and its theories and ideas; technology, meaning pieces about science applied to daily life, medicine, space travel, and more; and culture, specifically, the interactions of science with the arts and the media, and the practices of scientists that define a scientific culture.

The new articles in *Science Sketches* represent the same general threads, but here they appear under different aspects of what each thread means. One influence came from another book I wrote in the same time frame as *Real Scientists*. In researching *Physics: A Very Short Introduction* (Oxford University Press, 2019), I was reminded that the word "physics" comes from a Greek root that means "nature." At its most fundamental, science is the study of nature. Along with art and religion, it is one way that humanity tries to make sense of the natural world and the universe that surround us. Thus one major heading in this book is "Looking at Nature Without and Within," an appreciation of how we have found ways to peer deeply into nature, and also of how much is left to be examined and understood.

Another influence has been my deepening understanding of the role of technology and applied science in our society in areas such as biomedicine, algorithmic decision making, and artificial intelligence (AI). Recent dramatic, society-changing events such as the successful development of COVID-19 vaccines, the growing realization of how police treat people of color in the United States, the daily impact of social media and surveillance, and our increasing appreciation of the lack of equity between women and men in science and elsewhere— all these give new weight to the power of technology to change our lives for the better, or for the worse. What I've read, seen, and encountered about these issues has extended my thinking and my writing about technology and its ethical uses. This resulted in many of the pieces, especially recent ones, that I chose to appear here under the heading "Technology in Society."

Some of this coverage also appears under "Science, Fiction, and Art," because of how the pandemic affected my writing and film viewing. COVID-19 directly inspired the pieces under "Imagining the Pandemic." The disease also redirected my writing about science in films, which had focused on Hollywood's science fiction extravaganzas. When the pandemic emptied movie theaters, the production and distribution of major feature films halted or diminished (with long-term results yet to be seen). Like many, I turned to my computer to watch films, including number of documentaries and stories from independent filmmakers. Many of these brilliantly illuminate how science and tech affect individuals and society. Most of my comments about independent films appear under "Science, Fiction, and Art," along with articles about feature-length science fiction films that represent science and scientists well, or say something important.

This book presents different ways for you, the reader, to engage with science and technology. You can dip into the list anywhere to choose a short appetizer or a long entrée. For many of the articles, you can read further in the set of references I provide. But the whole menu is here for you to enjoy. I hope you will.

Sidney Perkowitz
Atlanta, GA, and Seattle, WA, USA
2020–2021

Acknowledgment

I'm happy to acknowledge the efforts of Jenny Rompas and Stanford Chong of Pan Stanford Publishing, who initially approached me about the possibility of writing a book for them. They liked my idea of an anthology of my writing. Jenny published my first anthology *Real Scientists Don't Wear Ties* under the new imprint Jenny Stanford Publishing, and she and her staff proved a pleasure to work with. After that good experience, I'm delighted to continue my interaction with the same group of people to produce this new collection, *Science Sketches*.

List of Illustrations

The illustrations are still images taken from the films listed below, courtesy of the Everett Collection with additional credits as given.

Looking at Nature Without and Within

Introduction

Some scientists work within what we typically call "nature;" botanists and zoologists trek through havens for plant and animal life, and oceanographers study oceans while afloat in them. But nature is bigger and more complex than that, so other scientists do their research at the giant CERN particle accelerator, or the immense telescope in Chile's Atacama Desert, or in their own labs full of exotic equipment. These are tools to study nature, that is to say the universe, in all its parts. We too are part of nature, and so other scientists study humanity individually or in groups using tools appropriate for biomedicine, neuroscience, sociology, and the other human sciences.

Science always seeks to look more deeply than our eyes can see. Our visual apparatus is remarkably acute but what it sees is a mere fraction of what there to sense in the universe, whose natural processes produce radiation extending from gamma rays and X-rays to radio waves. As the pieces below illustrate, these wavelengths can also penetrate and probe nature's small and large scales and where it is opaque.

Science Sketches: The Universe from Different Angles
Sidney Perkowitz
Copyright © 2022 Jenny Stanford Publishing Pte. Ltd.
ISBN 978-981-4877-94-7 (Hardcover), 978-1-003-27496-4 (eBook)
www.jennystanford.com

The invisible universe

"How to See the Invisible Universe" (2020) discusses how invisible long-wavelength radiation from space allows scientists to examine the early history of the universe, take a close-up view of a black hole, and probe the origins of life. These cosmic wavelengths were accidentally found in 1964 during engineering measurements at Bell Telephone Labs in New Jersey. "Heat Wave" (1991) describes how the great English astronomer William Herschel followed up his observation that certain colors seen through his telescope seemed to carry more heat than others. In 1800, he discovered "heat waves," the invisible infrared cosmic light at wavelengths beyond the red end of the visible spectrum. After describing Herschel's astronomical work, the article goes on to explain the broadly important uses of infrared radiation on Earth.

At the other end of the visible spectrum, "Mood Indigo" (1993) relates how in 1801 Herschel's work inspired the German scientist Johann Ritter to to seek and find invisible short wavelength ultraviolet light beyond violet light; and how in 1895 Wilhelm Roentgen accidentally discovered X-rays at even shorter wavelengths that gave them great penetrating power. These invisible wavelengths enhance medical practice and, when generated at high intensities by gigantic particle accelerators, contribute as well to other areas such as studying molecules and materials.

Black holes, quanta, and gravity

The preceding article "How to See the Invisible Universe" described a scientific effort to photograph black holes. This interest goes deep. Three researchers earned the 2020 Nobel Prize in Physics for expanding the theory of black holes and establishing that a monstrous black hole exists at the center of our own galaxy. "A Supermassive Lens on the Constants of Nature" (2020) describes the research that produced this Nobel award and also shows that the galactic black hole provides a novel tool to study a particular constant of nature that defines the universe and perhaps the fact that life exists as it does. Black holes equally fascinate the general public as I've seen in my own outreach activities and as confirmed by the data I discuss in "The Most Popular Physics Meme Ever" (2015).

One reason that physicists are interested in black holes is that work by Stephen Hawking suggests that their behavior gives clues to how to merge the two great theories of physics, quantum mechanics and general relativity. But unifying these theories has been an elusive goal. "Can Space Experiments Solve the Puzzle of Quantum Gravity?" (2020) explains why this is difficult, and describes new experiments in space that may give answers. The difficulty is partly due to the fact that over a century after Max Planck introduced the quantum theory, we still hardly understand it. That puzzlement is described in "Small Wonders" (1993). Today full understanding of the quantum still eludes us but "The Quantum Random Number Generator" (2019) shows that researchers have learned to use its peculiar properties in unique real-world applications.

Observing our Earth

The previous sections covered nature at its biggest and smallest scales. Here I consider the intermediate scale, which is what the word "nature" is usually taken to mean; that is, our own planet, its phenomena, and the life upon it.

"Sunlight, Life, and Time" (2018) addresses our Sun, the Earth's essential natural energy source. The piece describes four films from independent filmmakers about the violent solar activity that produces solar radiation and its effects on Earth, from beneficial photosynthesis to harm from ultraviolet light. The solar energy reaching us is critical in setting the Earth's temperature and the onset of global warming. One of the films also makes the point that whatever happens on the Sun, we know about it only 8 minutes later because of the finite speed of light.

"Flash!" (2019) is a scientific and cultural history of violent natural activity in our atmosphere, the phenomenon of lightning. Carrying meaning in ancient mythology and religion, a lightning bolt also carries tremendous power. This destructive capability is increasingly important as the effects of global warming make lightning strikes more prevalent around the world, but there is also the possibility of extracting useful power from lightning.

"We and the Earth Breathe Together"(2019) considers 10 science-based independent films that tackle climate change and the atmospheric pollution that encourages it. Some of the films paint a dystopic future where the atmosphere is so toxic that people must wear breathing apparatus or buy oxygen, but others hold out hope and give evidence that we may be able to turn back these effects. I treat climate change again later in the pieces "If Only 19th-Century America Had Listened to a Woman Scientist" and "Science Advances and Science Fiction Keeps Up."

Peering inside the body and the mind

In 1895 Wilhelm Roentgen's discovery of X-rays gave science and medicine the power to look inside the human body without cutting into it. "The Better to See You With" (2019) explains how it and other techniques, such as ultrasound and magnetic resonance imaging (MRI), were developed to give an array of tools to examine the body and the brain. Since the piece was written, there has been progress that allows MRI to be used diagnostically even on some people with implanted heart pacemakers and defibrillators, which was not initially possible.

One form of MRI, functional MRI (fMRI), makes it possible to examine the brain's neural processes as it forms thoughts; that is, to study the mind. One brain–mind interaction probed by fMRI is synesthesia, the mixed sensory response such as seeing images when hearing sounds that some people experience. "The Power of Crossed Brain Wires" (2020) gives the history and science of synesthesia, illustrated by my personal involvement with it. The piece also shows how synesthesia may give clues to an enduring mystery of neuroscience, how physical brain activity produces our own sense of self-consciousness.

We also know that the brain's neural connections are electrical in nature. "Can Zapping Your Brain Really Make You Smarter?" (2019) gives a brief history of how this knowledge has grown since the early Greeks, and how it is used today in transcranial direct current stimulation (tDCS) in which a small electrical current is applied to the brain. Some research shows neural benefits from tDCS, but scientists warn that performing tDCS at home with unregulated consumer devices may be unsafe.

How to See the Invisible Universe

In 1609, the great Renaissance scientist Galileo Galilei put a handheld telescope to his eye and looked to the heavens. In doing so, he opened the universe to direct human vision. Today, it remains a thrill to see Saturn's majestic rings through an optical telescope, as Galileo did. Astronomers and astrophysicists continue to learn about the universe, examining galaxies, stars, and planets at the visible light wavelengths.

Astrophysicists also study the invisible universe: at electromagnetic wavelengths, shorter than visible light; in the gamma ray and ultraviolet regions; and at even longer wavelengths, in the infrared. Each range gives new information. But it was a surprise when we found how much more information there is at still longer wavelengths, millimeters to centimeters. We generate such waves within microwave ovens and automotive cruise control systems. These waves also occur naturally in space, where they carry clues about the birth and growth of the universe, the centers of black holes, and the origins of life itself.

It's a truism in science that important discoveries often arose from serendipitous events. The German physicist Wilhelm Roentgen discovered X-rays after he saw an unexpected glow from a fluorescent screen in his lab. The French physicist Henri Becquerel discovered radioactivity when he noticed that photographic film stored in a drawer had become unaccountably fogged. Roentgen and Becquerel won Nobel Prizes in Physics for their discoveries. These researchers displayed the observational skills and the curiosity that lie at the heart of science, bringing us to a deeper understanding of nature.

Likewise, observations of the invisible universe, detected by means of long-wavelength photons in space, were first found by accident, in 1964. In a project to develop orbiting communications satellites, researchers Arno Penzias and Robert Wilson, at Bell Telephone's New Jersey laboratories, used a ground-based antenna pointed at the sky. Unexpectedly, it picked up a signal of unknown origin at a wavelength of 7.35 cm, which remained constant no matter where in the skies the antenna pointed.

To study this radiation without interference from the Earth's atmosphere, in 1989, NASA launched the Cosmic Background

Explorer (COBE) satellite into space, equipped with instruments to measure the strength and wavelength of millimeter and centimeter waves. The results, published in 1993, showed a distinctive peak at 1.07 mm, a "blackbody curve," which describes the electromagnetic waves emitted by any object above absolute zero temperature. The peak intensity and wavelength depend on the object's temperature. Our Sun, a hot body at 5,800 Kelvin (K), emits visible light with a peak at 500 nm. The COBE data perfectly followed the same theory but calculated for the extremely low temperature of 2.725 K, which generates millimeter and centimeter waves.

The COBE measurement of the so-called cosmic microwave background (CMB), along with the fact that the universe is expanding, provides strong evidence for the Big Bang. According to the theory, the universe began 13.8 billion years ago at an unimaginable density and a temperature of billions of degrees. The Big Bang produced highly energetic short wavelength photons that survive today as relics of cosmic birth, although they have changed: as the universe cooled and expanded, the photons carried lower energies at longer wavelengths. Today, they fill a universe whose temperature is near absolute zero. The clincher is that the measured value, 2.725 K, agrees with the theoretical prediction of 3 K based on the Big Bang—a prediction made in 1965, shortly after Penzias and Wilson made their accidental discovery.

The COBE results showed something else that ground-based data had not: the CMB—and therefore the temperature—was not perfectly even but varied slightly across the sky. This was important news about the state of the universe at the time when photons began traveling through it, 400,000 years after its birth. The temperature fluctuations track changes in the density of the hydrogen that then filled the universe. These density variations are the seeds that grew into today's cosmic macrostructure, consisting of strings of galaxies surrounding huge empty voids.

Mapping the Big Bang

To closely examine the density variations, in 2009, the European Space Agency (ESA) launched its Planck spacecraft, named after Max Planck, who derived blackbody theory in 1900. With improved technology (compared to COBE), the Planck spacecraft scanned the skies at nine wavelengths between 0.35 mm and 1 cm, measuring temperature differences down to 1 μK. After the spacecraft gathered the data, ESA scientists turned it into a high-resolution map of temperatures in the early universe as they appear in the CMB.

Scientists analyzed that map with the Big Bang theory and general relativity (Einstein's theory of gravitation) in mind. The goal was to see how the density variation produced a universe that contains all of the following:

- ordinary matter, the kind that surrounds us on Earth;
- dark matter, which exists in space and has gravitational effects but cannot be seen;
- and dark energy, which seems to fill all space and acts to expand the universe against gravity.

The final results, announced in 2018, give the most precise and complete description of the universe to date. We now know that it is approximately 13.8 billion years old, and is made of 4.9% normal matter, 26.6% dark matter, and 68.5% dark energy. To underline the point: 95.1% of the cosmos consists of entities unlike anything on Earth, whose nature we do not fully understand—we can only speculate until we learn more.

The analysis of the Planck data provided another surprise in the value it gave for a particular number, the Hubble constant, H_0. In 1929, the American astronomer Edwin Hubble, observing galaxies through what was then the world's biggest optical telescope, at Mt. Wilson, California, confirmed earlier ideas that the universe is expanding. Hubble derived a value for H_0, which gives the rate of expansion at different distances from the Earth or any other specific spot in space. H_0 has since been recalculated from newer astronomical data, but the value from the Planck data was 8% smaller than the recalculated value, indicating a slower expansion rate in the young universe. That discrepancy is now under intensive scrutiny, although we may have to wait until a planned new space mission, to take place in the mid-2020s, helps us to learn if it is due to an error or represents new knowledge.

Taking a picture of a black hole

Observations at millimeter wavelengths also made possible the first image of the most exotic cosmic object we know, a black hole. These regions, where incredibly dense matter produces a gravitational field so strong that not even light can escape, were predicted from general relativity. Since then, they have been observed in our own galaxy and elsewhere—not directly, but by means of gas molecules and dust particles pulled in by the powerful gravity. These components collide and generate tremendous heat, X-rays, and gamma rays, creating a glowing accretion disk around the hole.

In 2009, the Event Horizon Telescope (EHT) research consortium set out to image a black hole at the center of a distant galaxy denoted as M87. An event horizon is the imaginary surface around a black hole that represents the "point of no return;" once past it, no incoming object or photon can leave. But measurements had shown that photons of about 1 mm wavelength could escape the intense gravity just outside the event horizon and emerge through the accretion disk. EHT planned to detect these photons and turn them into a picture.

This was a tall order, one that required an array of ground-based radio telescope dishes to form an image. At a distance of 55 million light years, the target area within M87 appears as a tiny dot, about the size of a U. S. quarter viewed from 100,000 km away. To obtain an acceptable image, the researchers had to minimize diffraction, where electromagnetic waves are distorted as they enter an aperture, like the bowl of a radio telescope. The bigger the aperture, the less the diffraction. EHT managed the diffraction with a clever scheme that coordinated observations from eight different radio telescope installations around the world. This created an Earth-sized virtual telescope with extremely high resolution.

After recording and analyzing data measured at a wavelength of 1.3 mm, in April 2019, EHT presented its image. The by-now-familiar picture clearly shows a dark "shadow" inside the glowing accretion disk at the center of M87. The shadow closely surrounds the black hole's event horizon, making this the nearest we have come to pinpointing a black hole itself. The data shows that the mass within the black hole is 6.5 billion times that of our Sun. This supports what has been long surmised, that "supermassive" black holes lie at the center of galaxies, where they produce accretion disks called quasars, the brightest known astronomical objects.

Imaging the beginnings of life

Finally, perhaps the most intriguing use of long-wavelength radio astronomy seeks the beginnings of life in an inanimate universe. One theory for these origins is that the necessary complex molecules, such as the amino acids that form proteins, were created by chemical processes in space. These molecules then seeded life by coming to Earth and perhaps other planets. Judging by the life on Earth, the needed compounds are invariably organic molecules, containing carbon, hydrogen, oxygen, and nitrogen. Organic molecules and amino acids have been found in meteorites that reached Earth, which inspires the search for them in space.

A molecule in space is identified by finding features at characteristic "fingerprint" wavelengths in the radiation the molecule emits or absorbs. The organic molecules important for life processes are relatively massive, with ten or more atoms, and typically produce fingerprint features at millimeter to centimeter wavelengths. Radio telescopes operating in this range have found many organic molecules among the more than 200 types discovered in our galaxy and elsewhere, containing up to 13 atoms. A measurement in 2003 reportedly detected an amino acid, but that has not been replicated since. However, "precursor" molecules have been found: molecules that could change into sugars or amino acids with just a few chemical steps.

One telescope system that was used in EHT is also highly effective in seeking organic molecules. In the Atacama Large Millimeter/ submillimeter Array (ALMA), 66 dishes working together form the world's biggest single radio telescope installation.

I visited the Atacama Desert, in Chile, years ago and remember a bleak environment—hardly an advertisement for the lushness of earthly life. But Atacama's altitude and dryness are ideal for the ground-based spectroscopic search for the molecules of life. In 2014, researchers using ALMA at 3 mm wavelength found isopropyl cyanide, the first organic molecule discovered in space with carbon atoms arranged like those in the amino acids contained in meteorites. It occurred within a giant cloud of gas and dust in our own galaxy, where new stars form. As the quest for complex, biologically significant molecules continues, researchers should point their telescopes to regions in space where stars and planets are in the process of being born.

References

These references can be found at their individual sources and in the JSTOR digital archive at https://www.jstor.org/.

C. L. Bennett et al., Scientific results from the cosmic background explorer (COBE), *Proceedings of the National Academy of Sciences of the United States of America*, Vol. 90, No. 11 (Jun. 1, 1993), pp. 4766–4773.

D. Psaltis and S. S. Doeleman, The black hole test, *Scientific American*, Vol. 313, No. 3, Special issue, 100 Years of General Relativity, Sept. 2015, pp. 74–79.

James G. Lawless et al., Amino acids indigenous to the Murray meteorite, *Science*, Vol. 173, No. 3997 (Aug. 13, 1971), pp. 626–627

R. Bustos et al., Parque astronómico de atacama: an ideal site for millimeter, submillimeter, and mid-infrared astronomy, *Publications of the Astronomical Society of the Pacific*, Vol. 126, No. 946 (December 2014), pp. 1126–1132.

Heat Wave

Sir Frederick William Herschel, the 18th-century English astronomer, was not one to let his curiosity go unsatisfied. His first love had been music, but after a few years of playing oboe in a German military band, he had grown restless. He found a greater challenge in music theory and symphonic composition. Enchanted with the mathematics underlying harmony, he was soon reading texts on pure mathematics. When that subject drew his attention to astronomy, he wrote to his brother: "It is a pity that music is not a hundred times more difficult as a science." By the time he reached his thirties, Herschel had embarked on an obsessive quest to build progressively larger telescopes while supporting himself as a chapel organist. And in 1781, at the age of forty-three, he discovered the planet Uranus with one of his homemade telescopes and became a celebrated astronomer overnight.

It is little cause for wonder, then, that many years later the same intensely curious mind would make a discovery of an entirely different kind—and arguably of greater significance than the one that brought him fame. It arose out of a subtle anomaly Herschel noticed in his daily study of sunspots. Because he had slightly damaged one of his eyes while observing the Sun, he had begun to experiment with dark glasses of various color. Endowed with what to most people would seem a heightened awareness of the ordinary, he was perplexed when he felt the Sun's heat more strongly with some colors than with others. So, in 1800 he embarked on an experiment to measure the temperatures of the different colors of light.

He set a prism in a window exposed to direct sunlight, then projected the spectrum of colored light streaming out of the prism onto the top of a table. On the table he placed two thermometers—one resting in the colored light, and the other, an experimental control, outside the projected light beam. And indeed, he found that the temperature registered by the thermometer exposed to the light increased steadily as the thermometer moved from the violet through the blue and the yellow to the red. But he also discovered something much more astounding: when he continued moving the thermometer beyond the red light, the temperature jumped sharply, whereas at the other end of the spectrum, next to the violet light, there was no temperature rise at all. He concluded that there are invisible heat rays residing next to the red.

Herschel had discovered infrared radiation, a vast empire of light that accounts for roughly half the Sun's energy output. If our eyes and minds responded to the immense variety of infrared radiation as they do to visible light, our visual universe would expand in richness, texture, and detail in ways we can only dimly imagine. Sight, of course, is critical to our lives, yet visible light is a small pond in the sea of invisible infrared radiation that surrounds us. More commonly experienced as radiant heat, infrared light is the primary means whereby objects at ordinary temperatures transfer their energy through empty space.

Twentieth-century science has put the infrared spectrum to work in a number of ways. Soldiers find targets in the dark with sensors that "see" in the infrared, as they demonstrated dramatically in the Gulf War. Herschel's own profession, astronomy, has lately been able to peer deeply and continuously into the infrared sky, through satellite instruments and high-altitude telescopes, to gather information about matter too cold and dark to shine in the visible spectrum. Perhaps most significant is that infrared radiation serves as a sensitive internal probe of the workings of complex new materials. For example, it has helped shape the development and fabrication of semiconductor materials, which lie at the heart of all electronic devices. Furthermore, it may be the diagnostic key for understanding how the phenomenon of superconductivity persists at relatively high temperatures in recently discovered families of ceramic-like materials. Infrared radiation could well be the light that guides solid-state physicists to a new wave of superconducting technology.

All the fundamental properties of infrared light follow from the insights of the 19-century Scottish mathematical physicist James Clerk Maxwell. Maxwell wrote down four equations that describe a phenomenon known as an electromagnetic wave, which always propagates at the speed of light, 186,000 miles a second. It turns out experimentally that all forms of light, including the visible and the infrared, are electromagnetic waves.

The most vivid image for understanding the nature of electromagnetic waves was devised by the 19-century English physicist and chemist Michael Faraday. Imagine positive and negative charges at rest at various points in space. Just as one feels the effects of static electricity on a dry winter day, the charges "feel" attraction toward or repulsion from one another across space. Faraday made the interaction visible by drawing lines along the direction of the forces joining the charges. The picture that results, the electric field, is a spidery web of lines of force, each beginning at one charge and ending on another.

It was Faraday's keen wit to grant physical reality to the lines, as if each one were a stretched rubber hand. If one of the charges in the assembly slowly starts to move, the lines joining it to the other charges can stretch or contract, maintaining their smooth shapes. But if a charge sharply changes speed, its connecting bands can accommodate to the change only by developing kinks. And, just like the displacement on a plucked guitar string or the curve in the thong of a cracked whip, each kink travels as a disturbance along the lines of force, finally to jostle the electric charge anchoring the lines at the other end. The original accelerated charge also constitutes an electric current as it moves, which gives rise to magnetic fields with their own lines of force. Hence a magnetic whip cracks as well, to make a magnetic kink along with the electric one. It is worth mentioning that the kinks generally travel much faster than the charges do, just as the disturbance in a taut string can move faster than the finger that plucks the string. In fact, an electromagnetic wave is nothing more or less than a kink in the lines of force, and, as Maxwell showed, it propagates from one charge to another at the speed of light.

Each crack of the whip, or momentary acceleration of the first charge, gives rise to a single electromagnetic disturbance, similar to the single flurry when a stone is dropped into water. But if the originating charge constantly changes velocity by oscillating about a fixed position, a steady train of waves results, the waves radiating outward like concentric ripples in a pond. And like the water waves, electromagnetic waves have a frequency and a wavelength, the distance from the crest of one wave to the crest of the next. Since the waves move at the speed of light, the number of waves that pass a given point in one second, multiplied by the length of each wave, must be equal to a distance of 186,000 miles. Thus, if the wavelengths are substantially shorter than that enormous distance, a great many of them must cross a given point in a second, or in other words the frequency of the wave oscillations must be high. In any event, if you know the wavelength, you can calculate the frequency of the oscillation, and vice versa. The only property that really distinguishes among electromagnetic waves is their wavelength, or equivalently, their frequency.

Infrared light ranges in wavelength from 750 to 1,000,000 nanometers. (A nanometer is a billionth of a meter.) The range is further subdivided into the near (750 to about 10,000 nanometers), the mid (10,000 to 50,000 nanometers), and the far. Beyond that lie even longer waves—millimeter waves and microwaves. The waves just shorter than the shortest infrared waves are the ones the eye registers as red, and the visible spectrum then continues through progressively shorter wavelengths to the violet, with a wavelength of 400 nanometers.

In fact, the diverse species of light can be thought of as stretched out along a continuous spectrum of wavelengths, similar in concept to the row of keys on a piano. Each key differs in the frequency of the note it sounds, and notes one octave apart differ in frequency by a factor of two, The A above middle C has a frequency of 440 hertz, or 440 waves a second, and the A one octave higher has a frequency of 880 hertz. The wavelengths of visible light thus encompass less than one full octave—only eight white keys on the piano—whereas the invisible infrared world discovered by Herschel covers more than ten octaves—an entire keyboard full of light, and then some.

This picture makes it possible to understand why radiation at infrared wavelengths is so bound up with our everyday lives. Any matter at a temperature above absolute zero radiates electromagnetic waves to one degree or another simply because the atoms within the matter vibrate. Vibrating atoms are the manifestation of temperature; as the temperature rises, they vibrate faster. Because each atom has an overall electric charge that oscillates along with it, the atoms generate electromagnetic waves at the frequency of their oscillations. At the temperatures encountered in ordinary experience, the frequencies of the waves generated by the random, thermal oscillations of atoms lie in the infrared, and mainly in the near and mid ranges.

Thus, every ordinary object—a tree, the human body, a military tank—acts as a miniature sun, giving off radiant heat to its surroundings. Detecting those emissions can give night vision to a gunner or enable an earth-orbiting satellite to make the distinction between vegetable and mineral. But by the same token, ordinary objects can also absorb electromagnetic waves at infrared frequencies; as the incoming waves of infrared radiation impinge on the atoms, the atoms vibrate faster, and when such energy is absorbed, the absorbing object warms up. Anyone who has moved from shade to sun can testify to the instant feeling of warmth from infrared radiation streaming through the atmosphere at the speed of light to the skin.

It is partly its capacity to interact with atoms at or near the frequencies of their thermal oscillations that makes infrared light so useful for analyzing solid materials. Beyond the short end of the infrared spectrum, the wavelengths of visible light are too short to have much effect on vibrating atoms, and the wavelengths beyond the other end of the infrared spectrum are too long.

But to gain a more complete picture of infrared interactions in a solid, it may be useful to think about an alternative model of light that is put forward by quantum mechanics. In that model electromagnetic radiation can also be viewed as a packet of waves called a photon. A photon is a bundle of energy with no mass, which in some contexts acts as a wave and in other contexts as a particle. The amount of energy carried by a photon is directly proportional to the frequency of its light: the higher the frequency—which is equivalent to saying the shorter the wavelength—the higher the energy of the photon. Hence photon energies along the electromagnetic spectrum increase smoothly from the relatively low frequency infrared photons to the high-frequency X-ray ones. Like a hard-thrown rock that shatters glass, an X-ray photon is so energetic that it can break the chemical bonds holding things together. Infrared photons are more like Wiffle balls, bouncing harmlessly off the chemical bonds.

The upshot is that infrared photons have wavelengths small enough to penetrate atomic structures, but they are not so energetic that they do permanent damage. They are the gentle probes of the atomic world, altering perceptibly in response to every feature in the solid terrain they encounter. As infrared light travels through space, it resembles a deep ocean wave moving unencumbered over vast ocean spaces. The swell rolls on for mile after mile with little alteration, but it abruptly changes height and speed when it meets a new medium—a sandy beach or a rocky shore. When an infrared wave meets matter, it also encounters a new territory, rich in electric charges—electrons, positively charged atoms with missing electrons, and negatively charged atoms with extra electrons. As the charges pick up energy from the infrared electric field, the wave loses energy and changes its velocity. The combined effects of all the charges on the wave make it possible, at least in principle, to draw a detailed map of the microscopic arrangement of electrons and atoms.

In practice it turns out that making such a map would be far too complicated, were it not for the simplifying effects of the solid itself. In contrast with the amorphous arrangement of atoms in gases and liquids, the atoms in a solid—at least in most solids of importance to science and technology—are fixed in a precise pattern within a crystal lattice of chemical bonds. Like the framework of steel girders that makes up the skeleton of a skyscraper, the atomic pattern is repeated many times, though it gets interrupted occasionally by defects or impurities. Investigators derive the most useful information about the lattice and its electrons by comparing the intensity of the infrared light shone onto a material with the intensity the material transmits or reflects; the difference between the input and output intensities is the energy the material itself has absorbed.

A solid can absorb energy in three main ways, each of which can tell workers something about its internal makeup. The first kind of absorption gives a probe of the electronic structure of the solid: Does it conduct electricity like a metal, insulate like a plastic, or possess the intermediate properties of a semiconductor? One must again resort to quantum mechanics to understand the differences.

Every atom in the crystal lattice carries a set of electrons. Like threads in a tapestry, some electrons are woven deep into the lattice, whereas others may rove about freely, ready to galvanize into an electric current the instant a voltage is applied. The close quarters of the lattice confine electrons to energy bands. A single atom, far removed from others, accommodates its electrons at discrete levels of energy, but when atoms coalesce in a solid, the electron energies lie in continuous bands. A further constraint on the electrons comes from the interplay of quantum mechanics and the strict regularity of the lattice. In quantum theory, all particles can also act as waves, and so, like the photon, the electron is both a particle and a wave. Furthermore, the energy of an electron depends on its wavelength, and some electron wavelengths do not fit comfortably within the dimensions of the lattice. Consequently, some energies may be excluded by the lattice itself, which leads to gaps in the energy bands. The bands of allowed energies and the gaps between them are known as the band structure of the solid; if electron energies are represented as increasing along a vertical axis, you can imagine the energy bands and the forbidden gaps between them as a series of alternating horizontal stripes.

Now imagine an incoming photon that carries a discrete packet of energy associated with its wavelength. The photon can be absorbed by electrons in the solid only if the photon's energy is large enough to promote an electron to the next empty level of higher energy within the band structure of the solid. The most energetic electron in the solid is the one that gets promoted if the photon is absorbed at all. Such an electron always sets its sights for life on the open road—the independent life of the free electron in the solid.

In a metal, the most energetic electrons reach only partway to the top of an energy band, far below the nearest band gap, and so the photon is readily absorbed. Physically such electrons are so loosely associated with individual atoms in the lattice that the slightest energy makes them gain velocity and roam freely about in the solid. Such mobile electrons are also the ones that would carry an electric current in the metal, should a voltage be applied.

But suppose a large energy gap looms just above the most energetic electron. Such an electron finds itself imprisoned in a sedentary life most of the time; only a great burst of energy can help it vault the band gap to freedom. That is precisely what takes place in an insulator; the band gaps are so large that even great injections of energy lead to paltry currents. A semiconductor has smaller band gaps, and so it conducts electricity much more readily than an insulator, though not as well as a metal. In fact, at any reasonably warm temperature—warmer, that is, than a few degrees above absolute zero—the random jostling of the electrons by heat energy alone is enough to promote a few of them above the gap, and so a semiconductor does conduct small currents at room temperature. It should be noted that there are not enough free electrons in a pure semiconductor alone to make a useful device; instead a dopant, or impurity, is generally added. The dopant allows more electrons to hop up into the electrical freedom of the next higher band, thereby enabling the flow of a significant electric current.

Such freely moving electrons can absorb energy via a second mechanism: their negative charge makes them oscillate with the electromagnetic waves that make up the impinging light. An electron subject to such a force slips through a solid much as a wind-driven sailboat moves through water. The boat's energy is constantly dissipated by the frictional force of the water, which increases with the speed of the boat. Similarly, an electron slipping through a solid is opposed by a force that increases with its velocity, the net effect of its many collisions with other particles in the solid.

The third way solids absorb electromagnetic energy gives yet another sensitive probe of atomic structures. Impinging infrared waves drive the atoms that make up a crystal just as they do its electrons. In contrast with the electrons, however, the atoms are held in place relative to other atoms by bonds that act as microscopic springs. As an atom moves to the right, its neighbors pull it to the left, and vice versa, causing it to oscillate. The vibrations of atoms throughout a crystal send great shudders through the lattice structure, similar to the complex vibrations of a bridge buffeted by winds and traffic. The so-called resonant modes of the lattice—the vibrations whose frequency and geometry absorb the most energy—depend on the atomic masses in the crystal and the forces that link them. In quantum mechanics, the lattice modes are described as phonons, or bundles of mechanical energy.

Some phonons are actually waves of compression and rarefaction of the lattice bonds, much like sound waves in air, and at comparable frequencies. But most semiconductors also sustain more complicated modes of phonon vibration, whose frequencies lie in the much higher range of mid-infrared and far-infrared light. When the frequency of light impinging on a material matches the phonon frequencies, the maximum transfer of energy takes place. Thus, the phonon frequencies of a crystal can be determined by testing the material with various infrared frequencies to see which ones are more readily absorbed.

Infrared light has served as a fast, versatile, and nondestructive way of characterizing semiconductors ever since the invention of the transistor in 1948. There are many kinds of semiconductor, each with its own unique band gap and characteristic phonon frequencies. One of the simplest, elemental silicon, is the heart of the integrated-circuit industry.

It happens that many semiconductor band gaps are just the right size for infrared photons to kick electrons across the gaps. One can determine the width of the gap, or in other words the energy needed to promote an electron across it, by measuring the absorption of infrared light. Once the electrons are freed, their continued absorption of energy yields information about their density and speed: both measurements are good indexes of the solid's electrical properties.

Many newly fabricated semiconductors are binary, ternary, or quaternary compounds, made up of two, three, or four (or more) elements. In principle, the absorbed energy corresponding to phonon frequencies can signal the proportion of each element in the lattice. Each pairing of different kinds of atom generates a unique phonon frequency in the infrared spectrum. The phonon behavior of binary materials such as gallium arsenide is well established, but the corresponding study of ternary and quaternary materials is much less advanced.

The utility of infrared light as a probe is not limited, however, to the ways it can be absorbed. Whenever light crosses into a new substance, some of it also gets reflected. Exploiting that property has become more important since the semiconductor industry has developed an appetite for single and multiple films only micrometers thick. (A micrometer is a millionth of a meter.) Structures made up of hundreds of layers whose composition alternates between one kind of semiconductor and another have become the state of the art. Because they mimic the repetitive geometry of natural crystals, the structures are called superlattices. With the proper choice of materials and layer thicknesses, a superlattice having virtually any band gap can be constructed.

What good is infrared light as a probe of those layers? Visible light is readily absorbed by atoms, and so any reflected light from a material comes from near the surface, within the first 100 atoms or so. But because infrared wavelengths are less readily absorbed, they can bore some 100 to 1000 times deeper into the material. When an infrared light wave reaches the second layer from the top of the semiconductor, it is reflected at the interface and races back to join the wave reflected at the top layer. The extra distance traveled by the second wave offsets its peaks and troughs from those of the first wave. The offset gives a precise measure of the thickness of the top layer. Because light is reflected similarly at every other boundary between the layers of the electronic cake, the thickness of each layer can be measured in the same way. In 1986 infrared radiation probed a superlattice for the first time: 170 layers each of mercury telluride seven nanometers thick alternating with cadmium telluride four nanometers thick.

The low-energy photons from the far infrared may be able to shed light on another kind of energy gap found in the astonishing materials known as superconductors. When a material becomes a superconductor, it seems to defy the laws of physics: a closed loop of superconducting material will carry current indefinitely, without any loss of energy, even after the battery that initiated the current has been removed. (Perpetual motion is precluded, however, because friction sets in whenever useful power is extracted.) Since their discovery in 1911, superconductors have teased the imaginations of scientists with applications of a wonderful promise: incredibly powerful motors and magnets, levitating trains, high-speed computer chips.

But in spite of tremendous efforts, a practical superconductor remains elusive. The vexing obstacle for many years was the need for extreme cold; the materials had to be cooled almost to absolute zero before they would become superconducting. Then, in 1986, Johann Georg Bednorz and Karl Alex Muller, both of the IBM Research Laboratory near Zurich, startled the world with the discovery of a new class of ceramic-like materials that become superconducting at temperatures high enough for practicality—some as high as 100°C warmer than the cutoff temperatures for previously known materials. Nevertheless, the new superconductors present their own difficulties, and many workers believe the difficulties will be overcome only after theorists understand how superconductivity arises in the new materials.

A successful theory of cold, metallic superconductors was proposed in 1957 by the physicists John Bardeen, Leon Cooper, and J. Robert Schrieffer, who all were then at the University of Illinois at Urbana. According to the BCS theory (for Bardeen, Cooper and Schrieffer), the free electrons in a superconductor do not repel one another, but attract instead, albeit weakly. As an electron moves through a lattice of atoms, its negative charge attracts the positive charges around it and thereby distorts the lattice. The positive charges in the distorted lattice form a kind of shell around the electron, and that shell attracts a second electron. The two electrons form a weakly bound pair, which has deep quantum–mechanical implications.

The rules of quantum mechanics allow the bound pairs—known as Cooper pairs—to circulate through the atomic lattice at the same low-energy level. In that state they move easily in one large mass when a voltage is applied. The movement continues unperturbed by outside agents unless the pairs break up. Only a minute amount of energy initiates divorce, which is why superconductors work only at low temperatures. The pair-binding energy is another kind of energy gap, and the BCS theory predicted it to be in the range of the extreme far infrared.

The prediction proved triumphantly correct for the metallic superconductors. But the new superconductors are so different from the earlier ones that many physicists think they must be described by a new, as yet undetermined theory. If they do work according to the old theory, experimenters should find an infrared wavelength that effectively cancels the superconductivity, presumably by breaking up Cooper pairs. Workers are still struggling to make consistent measurements of such an energy gap because of the complicated technology needed to generate longer infrared wavelengths of sufficient intensity.

Investigators have been forced to invent clever ways of harnessing enough mid- and far-infrared radiation to probe materials. The most obvious method is to heat a filament and let its emanations bathe the solid to be probed. But there is a catch-22: The most prevalent wavelengths are radiated in the far infrared only by relatively cold filaments that give off little power. Yet the hotter a filament gets, the shorter its most prevalent radiated wavelengths become, until they range out of the infrared region. Even a lamp operating at a temperature of 700°C and 100 watts of power emits only 100 microwatts of power in the far infrared.

Furthermore, if a measurement is to be useful, the effect of each wavelength must be analyzed separately. For some time the instrument of choice has been the interferometer, invented in 1881 by the American physicist Albert A. Michelson. More recently other techniques have come to the fore. Gas lasers such as the carbon dioxide laser emit high power in the infrared but only at highly select wavelengths. Two newer instruments, the synchrotron and the free-electron laser, give broader coverage and higher power—as much as several megawatts in brief pulses. But because they are based on particle accelerator technology, they are large and expensive—the synchrotron is a gargantuan machine of the size of a football field. Infrared free-electron lasers are now operating at the Los Alamos National Laboratory, the University of California at Santa Barbara and elsewhere.

Recently my colleagues from Emory University and from the National Synchrotron Light Source at the Brookhaven National Laboratory on Long Island and I made the first measurement of infrared transmission through a thick film of a high-temperature superconductor. The measurements suggested that BCS theory does apply, but the infrared spectrum is so complex that no definitive ruling can yet be made.

The infrared analysis of materials is deeply tied to their use, and few such uses of science are untainted. Infrared technology began in the Second World War with military applications. Today a laser saves a life in surgery and takes a second life in war; a computer chip, built with the knowledge that infrared analysis affords, educates a student or guides a missile.

But a world without infrared technologies would be a sadder one indeed. As I listen to *La Boheme* or Mozart from a compact disc, I think of the minute infrared laser inside the player, flawlessly reading the bit-encoded music captured in the disc, and of the silicon chips that shape and enlarge the cold binary signal into warm but crystalline sound. As 1 leave my office at night to gaze up at the cold winter sky, I am reminded that its infrared emissions, invisible to me, recently made a cooler, hidden universe available to astronomers.

I like to imagine that William Herschel, astronomer and musician, would be pleased that his scrap of invisible light widened our view of the universe and brought more music into the world. I think he would smile as I explain how his infrared waves now even read notes and help fashion the electronic devices that bring the music to our ears. Herschel found infrared light in the Sun's celestial sphere. Two centuries later those periodic waves are transformed by a compact disc into a sound sublime. Pythagoras of Samos, among the first to meditate on the celestial harmonies, might agree with me that Sir William finally brought to earth the music of the spheres.

Mood Indigo

On travels through Bavaria some time ago, I had the good fortune to visit the University of Würzburg, nestled amid the rolling wine country along the river Main. I am a physicist, and so, though I travel primarily to enjoy the simple pleasures, I was thrilled when my hosts recognized my professional interests and put me up in the very room once occupied by Wilhelm Conrad Roentgen. If memory serves, I even slept in Professor Roentgen's own bed. Or perhaps I only dreamed that, after an evening spent sampling the local wine. But there is no doubt that in 1895 Roentgen was living in Würzburg when he made a contribution to science that changed the world: the discovery of X-rays. That wonderful discovery helped spark 20th-century revolutions in medicine and physics, and it strongly contributed to the puzzling quantum theory of electromagnetic radiation. Roentgen's exploration focused attention on the invisible, powerful part of the spectrum extending from ultraviolet to X radiation and on to gamma and cosmic rays. The importance of the research was quickly recognized: in 1901, he was awarded the first Nobel Prize in Physics.

Roentgen was not the only explorer of the spectrum to begin his trek in Germany, where the wine or the air must stir some special creative element. How else can one explain that the discovery of ultraviolet radiation, which is spectrally adjacent to Roentgen's X-rays, took place less than a hundred miles northeast of Würzburg? In 1801, Johann Wilhelm Ritter, then 25 years old, was a respected scientist at the University of Jena and a newly appointed member of the court of the Duke of Gotha. Ritter had strong ideas about unity and polarity as principles of nature, exemplified in the linked but opposing north and south poles of magnetism.

Inspired by William Herschel's discovery in 1800 that invisible infrared rays lie beyond the red in the spectrum of sunlight, Ritter sought a polar twin in the form of unseen radiation beyond the opposite, violet end of the spectrum. His probe was a piece of paper soaked in a solution of silver chloride. It was already known that the preparation, a kind of proto-photographic film, turns black under visible light. But as he exposed the paper to the rainbow of colors emerging from a prism, Ritter watched the paper darken even more rapidly just beyond the violet portion of the spectrum than it did in violet light. Unseen rays appeared to be changing the silver chloride to silver.

Ritter did not know what caused the fade to black. And Roentgen did not understand the fearsome effect that enabled him to thrill the world by displaying the bones of a living hand. But today physicists know that those pioneers had turned up two key pieces in a larger puzzle. Ultraviolet and X radiation both are part of a much broader, unified physical picture of electromagnetic waves; they both are forms of light. Moreover, with increased understanding of those forms of light have come increasingly powerful applications. Certainly, both ultraviolet and X radiation, in excess, are damaging to the body. But in moderation they serve as powerful diagnostic tools, and they can even heal. X-rays are also employed in the study of crystalline solids, which form the backbone of the electronics and computer industries. And in recent years it has become clear that faster, more powerful, and more precisely focused beams of ultraviolet and X radiation can serve important scientific and technological needs. In many cases, those wavelengths are the most useful means of delivering enormous bursts of energy to a small area in an extremely short time.

To understand better the unseen light in the ultraviolet and X-ray region of the electromagnetic spectrum, think again about the colors of the rainbow. Those colors change across the rainbow in an orderly sequence, which young students memorize with the aid of the fictitious name ROY G. BIV. The letters in the mnemonic correspond to the initials of the colors, in order of decreasing wavelength: red, orange, yellow, green, blue, indigo, and violet.

The eye cannot see light whose wavelength is shorter than the deep violet, about 380 nanometers (billionths of a meter) between the crests of successive waves. But as the wavelength gets shorter, more wave crests pass a given point in space each second, or in other words its frequency gets higher. And the higher the frequency, the more energetic the light is.

But what could it mean to say that light is more energetic? In quantum theory there are two complementary pictures of light, to be applied more or less as dictated by the needs of explanation. All light of a given wavelength is associated with a photon, or particle of light having a definite energy; the shorter the wavelength, the more energetic the photon—the "harder" and more penetrating it is. Radiation with wavelengths between 400 and 4 nanometers is said to lie in the ultraviolet region of the spectrum. That region is further divided into UVA (between 400 and 320 nanometers), UVB (between 320 and 280 nanometers), and UVC (below 280 nanometers, the most energetic ultraviolet radiation of all).

Still more energetic are the X-ray photons, whose longest wavelengths are defined to measure 30 nanometers; the X-ray region thus overlaps a part of the ultraviolet region. X-rays whose wavelengths are longer than 0.1 nanometer are called soft, and those with wavelengths shorter than 0.1 nanometer are called hard—again, because of their increased penetrating power. Like a high-powered rifle bullet, an X-ray photon can scar even a solid crystal, and it can do terrible damage to biological tissue by breaking the bonds between the atoms that form its molecules and by tearing away electrons from the atoms themselves, or ionizing them. The less energetic ultraviolet photons are more like pistol bullets, but there is growing evidence that they cause even more harm to the soft tissues of life than do X-rays.

Two principal natural processes generate ultraviolet and X-ray photons. The first is the so-called thermal emission of electromagnetic energy, from any body whose temperature is above absolute zero. The atoms that make up the body vibrate back and forth at many different frequencies. Because they are electrically charged, each vibration sets up an electromagnetic wave at the frequency of the vibration. The result is a continuous spectrum of radiant energy that is broadcast outward across the virtually continuous set of frequencies represented by the enormous number of atoms in the body.

The energy broadcast at each frequency, however, depends on the temperature of the body. Only extremely hot bodies generate ultraviolet and X-ray photons as a substantial fraction of their output. Even the Sun emits less than a tenth of its radiant energy at wavelengths shorter than 400 nanometers. Some of that short-wavelength radiation comes from the solar surface, whose temperature is roughly 6000 degrees Celsius. Hotter internal parts of the Sun contribute additional continuous short-wavelength energy, as well as emissions at specific wavelengths that show up as bright lines in the solar spectrum.

Such bright-line emissions come from the second process that generates ultraviolet and X-ray photons. Every atom is made up of a positive nucleus, to which negatively charged electrons are electrostatically bound. As quantum theory has it, the electrons are arrayed around the nucleus only in specific orbits or shells, each with a definite energy. The single electron in an undisturbed atom of hydrogen, for instance, resides in its innermost orbit, a fraction of a nanometer from the nucleus and therefore subject to a strong pull that holds it in a tight embrace. An electron in an outer shell is less compulsively held; it is more nearly a free body, with higher energy. If an inner electron gains energy from some outside source, it alights in one of those distant shells, but it soon returns to its initial site. As it does, it gives off a flash of light, a photon. The photon's energy, and therefore its wavelength, is determined by the difference between the higher and lower electronic energies.

The electronic energies are determined by the strength of the nuclear attraction, which is set by the count of protons in the nucleus. The electron in a hydrogen atom is held by only one central proton. As that electron returns to its innermost shell from any outer one, it emits ultraviolet light whose wavelength is at least 100 nanometers. A heavier atom, however, includes more nuclear protons, the combined charge of which pulls an electron far more intensely. The 29 protons in the copper nucleus, for instance, can give rise to energy differences from shell to shell nearly a thousand times greater than the ones in hydrogen. The photons emitted when electrons fall into the innermost orbit have wavelengths of less than a nanometer: they are X-ray photons. Any reasonably heavy element can produce X-rays in that way. The wavelengths are sharply defined and are characteristic of the emitting element, since they come from electronic transitions between exact energy levels.

Because large energies are needed to generate X-rays, they are not produced naturally on the earth except in some radioactive processes. In the sun, X-ray line emissions occur because the great thermal energy promotes electrons to higher shells. And other sources in the distant cosmos produce X-rays at unthinkable powers. Since the first extrasolar X-ray emitter, Sco X-1, was discovered in 1962, hundreds of such astronomical sources have been observed. Many have been associated with visible stellar objects: Sco X-1 is identified with a star in the constellation Scorpius, and another source, Tau X-1, is located in the Crab Nebula. Sco X-1 is so powerful that if its emissions could be gathered and stored for one second, they would meet all the energy needs of the United States for a billion years. People are safe from such terrible cauldrons of radiation because the emissions are greatly attenuated over the light-years they travel to reach the solar system.

It is the far less powerful ultraviolet light from the earth's own furnace, the Sun, that is a harmful feature of daily life. Fortunately, the small fraction of the Sun's energy emitted in the ultraviolet is further diluted by the atmosphere, especially by its layer of ozone. The ozone molecule, made up of three oxygen atoms, effectively absorbs ultraviolet radiation at wavelengths shorter than 295 nanometers. Unfortunately, the protection afforded by that airy armor is jeopardized by the current thinning of the ozone layer, attributed to chlorofluorocarbons released by human activity. But even with an undamaged ozone layer the ultraviolet light reaching the earth's surface is energetic enough to cause harm. Solar UVB light may well be an important cause of skin cancer, and the incidence of that form of the disease is rising as people are spending more time at the beach.

Ultraviolet photons can cause damage directly by breaking chemical bonds in DNA, the genetic material in cell nuclei that determines the growth and development of the cell. But the interaction of ultraviolet light with human biochemistry is a tangled skein of effects. Some threads in the tangle give rise to serious illness and death; others cure disease.

Statistical evidence inescapably links solar ultraviolet radiation to some skin cancers. An increase of 1% or 2% in UVB radiation increases the incidence of nonmelanoma forms of skin cancer by 2% to 4%. But the connection between the more serious melanoma cancers and the sun is not so well understood. One important factor seems to be whether or not a person has had exposures to ultraviolet rays strong enough to burn the skin; that may explain why indoor workers are more at risk for melanoma than those who spend their days outdoors.

Although the exact mechanism for damage in melanoma is unclear, the direct damage that turns ordinary cells cancerous seems to be only part of the story. The sites where melanomas erupt on the body are not always the most heavily exposed ones. The far-reaching implication is that the absorption of ultraviolet light may suppress the body's natural defenses against tumorous cells. It is as if the ultraviolet bullets were coated with a poison that inflicts a devastating second-stage effect; after damaging the DNA in cells, the radiation goes on to prevent the body from containing and healing early injuries.

And yet, as if to compensate for its own most widespread evil, ultraviolet light can also be exploited to treat some kinds of cancer. In a procedure known as photophoresis a patient's blood is removed, irradiated with ultraviolet light and returned to the body. The procedure relieves symptoms of the skin cancer called mycosis fungoides, and it may also increase survival rates, with none of the side effects associated with chemotherapy. Photophoresis is promising as well for certain forms of leukemia.

Ultraviolet light is also employed to treat certain diseases by activating photosensitive drugs. One therapeutic application is for psoriasis, the chronic appearance of scaly dry patches on the skin; because the treatment relies on long-wavelength ultraviolet light, it is known as psoriasis-UVA (PUVA) therapy. The UVA works together with the drug psoralen to slow the process that causes psoriasis lesions. Every year between 25,000 and 50,000 people receive PUVA therapy in the United States. Unfortunately, that bright thread in the tangle of human responses to ultraviolet light is interlaced with a worrisome dark strand. The same PUVA therapy that relieves psoriasis may activate full-blown AIDS in people infected with HIV. And ironically, many HIV-infected patients have skin conditions that make them candidates for PUVA therapy.

The human response to X-rays is similarly tangled. In large doses they have terrifying effects on tissue, and even as recently as the 1950s the cumulative effects of repeated doses went largely unrecognized. It was not uncommon at that time to find X-ray machines in shoe stores, where customers could pay to wiggle their toes and watch the movement of the bones in their feet. X-ray exposures are now kept to a minimum for patient and operator, but they remain a staple of medical practice. They are also exploited for their destructive powers to treat some cancers. Abnormal cells are more sensitive to damaging radiation than are ordinary ones, and so they can be selectively destroyed.

The medical role of X-rays has been enhanced in recent years by computerized tomography (CT) scans, which instead of a flat picture give a three-dimensional view of the inside of the body. To generate the image a computer-controlled mechanism moves the X-ray source in defined orbits around the long axis of the patient's body, and an X-ray photograph is made at each of many positions. The real power of the process comes from another computer, which assembles the scans into a three-dimensional representation. The computed image enables the radiologist at a computer-graphics terminal to rotate a sculptural picture of a living skull or any other part of the bone structure, much as Hamlet mused over Yorick's remains.

X-ray analysis in nonmedical applications has also become an essential tool for understanding the modern world. Electronics, metallurgy, and other materials-based technologies draw heavily on the science of crystalline solids, which aims to determine how the atoms in the solid are packed together in space. Some arrangements are as simple as regular stacks of cannonballs, such as the array of atoms in silicon, the material for electronic chips. Others challenge the visual imagination, such as the ornate placement of the constituent atoms in yttrium barium copper oxide, a new high-temperature superconductor. But simple or baroque, a common feature is that the atoms typically lie less than a nanometer apart. Even with unlimited magnification, ordinary visible light cannot discriminate the details of those structures. Only light with a wavelength comparable to the interatomic spacing—that is, X radiation—can discern the features.

Wavelength is a gauge of measurement because it defines how waves act when they encounter a solid structure. Picture a moored buoy bobbing up and down in a train of regular ocean waves. As each wave crest reaches the buoy, the obstacle disturbs the wave front. The disturbances fan out, and so some waves move off in a direction different from that of the incoming waves. The net effect of the buoy is the diffraction, or bending, of the waves, a property inherent in wave behavior.

Now add a second buoy near the first, and imagine that one of the marching waves reaches the two buoys simultaneously. Identical diffracted waves form at the buoys and crisscross in the region between them. If the distance from crest to crest—the wavelength— of the original marching waves is greater than the spacing between the buoys, the emerging diffracted waves are still much in step. An observer would be hard put to tell whether two buoys or only one buoy had disturbed the original wave front. If the wavelength is the same or less than the spacing between the buoys, however, crests as well as troughs appear near the obstacles. The result is a distinctive interference pattern, strengthened where crest matches crest and weakened where crest matches trough. An observer studying the disturbed wave front could clearly see that the marching waves had encountered two obstacles.

Stripped of the romantic imagery of ocean rollers and clanging buoys, the example simply shows that when light waves illuminate a structure, they cannot pick out, or resolve, geometric features smaller than the wavelength of the light itself. Infrared, visible, or ultraviolet light cannot probe the nanometer-size details of a crystal. Only X-rays can.

And so early in this century investigators began to apply X-ray diffraction to examine and catalogue crystalline solids. The geometries and dimensions determined by the new technique formed the basis for microscopic theories of solids, which take concrete form in today's technology. X-ray analysis remains a prime tool for examining the structure of materials such as high-temperature superconductors, "buckyballs" (the new geodesic dome-like carbon-based molecules named after the architect Buckminster Fuller), and intricate compounds with biological significance. The technology has given rise to some of the most accurate experimental measurements ever made.

Lamps and lasers provide the ultraviolet light, and high-voltage electron tubes provide the X-rays needed for most ordinary work. But it is virtually axiomatic in science that exploring the limits of nature demands the most advanced technology available. Some investigations require X-rays or ultraviolet light at extraordinary intensities; other efforts demand a source that can be flexibly tuned to different wavelengths; others still must have brief bursts of radiation.

To meet those needs the best answer to date has been to build sources that can push electrons to enormous accelerations. The accelerated charges generate X-ray and ultraviolet radiation at high fluxes. Even the standard X-ray tube depends on accelerated electrons: they are boiled off a wire filament and accelerated across the evacuated tube by a high voltage. When the electrons smash into a target on the other end of the tube, the collision releases X-rays. But there are practical limits on the X radiation available from the linear design of the tube and its reliance on high voltage alone. More clever schemes are more productive, but the means are immense.

The most useful new source of ultraviolet and X radiation is the synchrotron, an enormous device that traces its origins to the accelerators originally designed to study elementary particles. In particle research itself that line of machinery has culminated in the astonishing Superconducting Super Collider, a racetrack for subatomic particles, fifty-four miles around, that is under construction in Texas.* Machines designed as synchrotron light sources do not require such a gargantuan scale, but their size is still impressive. The National Synchrotron Light Source (NSLS) at the Brookhaven National Laboratory on Long Island, New York, the nearly completed Advanced Light Source at the Lawrence Berkeley Laboratory in California, and the European Synchrotron Radiation Source (F-SRF) being built at Grenoble near the French Alps each could comfortably house a couple of dirigibles or enclose most of a football field.

Synchrotrons rely on magnetic fields to accelerate energetic electrons, which then radiate light in vast quantities. The push or pull of a magnetic field on moving electrons is a fundamental property of electromagnetism. A current of electrons creates a magnetic field; conversely, an electron moving in a preexisting field is subject to a force at right angles to its direction of motion. In the synchrotron light source powerful magnetic fields force a cluster of electrons to circle an enormous evacuated track. The magnetic fields continuously push the stream of electrons inward, toward the center of the circle. According to Newton's laws of motion, where there is a push, there is an acceleration. And as the electrons accelerate toward the center, they radiate light. The same synchrotron mechanism, scaled up to stellar size, is thought to be responsible for the intense X radiation observed from cosmic sources such as Tau X-1.

The amount of light radiated by any synchrotron source depends on the speed of the accelerated charges. In the earthbound version of the synchrotron the electronic speed is increased by feeding bursts of carefully timed energy to the orbiting charges. The energy and speed of the particles grow with every burst, until each electron carries billions of volts of electrical potential and is moving at nearly the speed of light. The resultant electromagnetic radiation is typically millions of times stronger than the radiation from conventional sources, and it spans a wide range of wavelengths. The NSLS facility, for instance, generates radiation at wavelengths that vary across nine orders of magnitude, from 0.01 nanometer (hard X-rays) to a centimeter (in the microwave region of the spectrum).

In addition to the many wavelengths available from a synchrotron source, an extremely bright beam can be generated. Brightness is a measure of the density of photons in the beam, and high brightness implies an intense beam focused on a small cross-sectional area. In fact, the full power of the synchrotron can be brought to bear on a minute target; conventional X-ray sources cannot be so focused.

The focusing is important for several reasons. First, quantum theory suggests that solids take on valuable new properties when they come in small, so-called mesoscopic packages. Those properties will become increasingly important as new generations of electronic chips are formed on such small scales. Second, novel materials can be made initially only in minute quantities, and X-ray analysis must be able to observe the atomic arrangements across only a few micrometers. In a recent experiment X-ray analysis was carried out on a crystalline filament of bismuth only one-hundredth the thickness of a human hair. Finally, the tight focusing of the synchrotron is particularly important for the study of biological substances such as proteins, which are difficult to form into large crystals.

Another advantage of synchrotron radiation is that it is pulsed, like an enormous photographic strobe light. The light blinks on as each group of electrons circles the ring. Thus synchrotron light can make rapid X-ray snapshots that freeze the action in the microscopic world one frame at a time. That capability is critical for examining certain biological processes. To understand how an enzyme works, for instance, one must study the chemical action of the molecule in real time. Such a study is planned for the synchrotron. The electrons in that machine will circle the ring a million times every three seconds, which will make possible virtual home movies of enzyme reactions, with successive frames spaced as little as three-millionths of a second apart.

Perhaps the most advanced medical use of synchrotron light is transvenous coronary angiography, under development at the NSLS Medical Research Facility. The technique employs synchrotron X-rays to examine the coronary arteries, which carry oxygen-rich blood to the heart. When the arteries are choked with fatty plaque, heart disease ensues. In the standard method of examination, a dye containing iodine is injected directly into the coronary arteries to enhance the X-ray contrast. But the injection is dangerous, and physicians dare use the technique for only the most pressing cases. The risk is lower if the injection is made into a vein, but then the dye is diluted before it reaches the coronary arteries. Any blockage is hard to see under ordinary low-power X-rays. The synchrotron light can be selected for the wavelengths most effective for enhancing contrast, and it can be shined at sufficient power to give excellent images. The X-ray dose to the patient is comparable to that with the conventional method.

A visitor to a synchrotron light source—with its immense activity, suggesting that important science is under way—might feel that human control of ultraviolet and X-ray light is complete. The truth is more complicated and even paradoxical, for much is still unknown about this forbidding part of the spectrum. Those of us who study matter are proudly certain of our quantitative findings when X-rays probe inert or dead material, yet we are puzzled and fearful when ultraviolet light pierces living cells. The deep emotional gulf reflects the differing grasps of the physical and the biological worlds.

I begin to experience the paradox in a personal way as 1 recall my travels in Germany, where Roentgen and Ritter found the intense X-ray and ultraviolet photons. After visiting Würzburg, I turned west and a little north to Mainz, where I boarded a side-wheel steamer to sail down the river Rhine. I remember taking the sun, being showered with radiation that physically bade nothing but ill for my exposed skin. But there I enjoyed a half-hour of complete happiness, one of those rare confluences of perfections whose true weight is felt only years later. Sitting relaxed aboard the ship, I sipped Rhine wine and watched and vineyards glide by. Perhaps enough German wine and air would have made me a spectral explorer like Roentgen and Ritter. But something else stays with me in crystalline vividness. I see yet the sparkling orderliness of the steamer, the fascinating variety of the passengers, the pale gold of the wine, all bathed in the pure, clean brightness of that sunlight. The joy I felt came from the magic of the moment, and from an innocence that marked a younger me. Few enough moments since have brought me such euphoria. For many of us few moments carry such pure pleasure as the ones lived dangerously in the Sun's light.

---

*The enormous Superconducting Super Collider was never completed. The project was cancelled by the U.S. Congress in 1993.

A Supermassive Lens on the Constants of Nature

The 2020 Nobel Prize in Physics went to three researchers who confirmed that Einstein's general relativity predicts black holes, and established that the center of our own galaxy houses a supermassive black hole with the equivalent of 4 million suns packed into a relatively small space. Besides expanding our understanding of black holes, the strong gravitational field around the supermassive black hole is a lab to study nature under extreme conditions. Researchers, including one of the new Nobel Laureates, Andrea Ghez at UCLA, have measured how the intense gravity changes the fine structure constant, one of the constants of nature that defines the physical universe, and in this case, life within it. This research extends other ongoing efforts to understand the constants and whether they vary in space and time. The hope is to find clues to resolve issues in the Standard Model of elementary particles and in current cosmology.

Besides Ghez, the other Nobel Laureates honored in 2020 are Roger Penrose at Cambridge University, who deepened our theoretical understanding of black holes; and Reinhard Genzel, of the Max Planck Institute for Extraterrestrial Physics in Garching, Germany. Ghez and Genzel carried out parallel but separate observations and analysis that led each to deduce the presence of our galactic supermassive black hole. At 27,000 light-years away, obtaining good data required huge telescopes. Ghez worked with the Keck Observatory on Mauna Kea in Hawaii, and Genzel used the Very Large Telescope in Chile. Each researcher found that the motion of the stars they observed arose from an enormous mass at the center of the galaxy. They obtained the same value, 4 million times the mass of our Sun, in a region only as big as our solar system—definitive evidence of a supermassive black hole.

Ghez's research at Keck made her a co-author in a paper published this year, in which Aurélien Hees of the Paris Observatory and 13 international colleagues presented results for the fine structure constant near our galactic supermassive black hole. Remarkably, Ghez's Nobel Prize–winning results supporting this research combined today's theories and astronomical techniques with ideas dating back to Johannes Kepler and Isaac Newton to examine the motion of stars near the supermassive black hole. This is another example of Newton's insight about how science advances when he wrote in 1675, "If I have seen further it is by standing on the shoulders of giants."

German astronomer Kepler is one such giant who changed science when he presented his laws of planetary motion in 1609. He was the first to show that the planets do not orbit the Sun in divinely inspired perfect circles, as had been assumed. The orbits are ellipses with the Sun at a focus of the ellipse, one of the two points symmetrically offset from the center that define how to construct an ellipse. Kepler also found a mathematical relation between the size of a planetary orbit and how long it takes the planet to complete a circuit.

In 1687 Newton gave Kepler's laws a deeper, more coherent physical basis. Newton's law of gravitation, based on mutual attraction between bodies, showed that a celestial object in a closed orbit around a mass follows an elliptical path that depends on that mass. This result, which today is taught in introductory astronomy, is the heart of how Ghez found the mass of the supermassive black hole. Her years of careful observations precisely defined the elliptical paths of stars orbiting the galactic center; then she used Newton's theory to calculate the mass at the center (general relativity, which replaces Newton's law, predicts black holes but Newton's approach is sufficiently accurate for the stellar orbits around the supermassive black hole). Knowledge of these orbits would be crucial for measuring the fine structure constant in the strong gravity near the supermassive black hole. How that constant depends on gravity could be a clue to modifying the Standard Model or general relativity to deal with dark matter and dark energy, the two great puzzles of contemporary physics.

This particular examination fits into a bigger, long-term examination of the fundamental constants of nature, each of which tells us something about the scope or scale of our deepest theories. Along with other constants, the fine structure constant (denoted by the Greek letter α), appears in the Standard Model, the quantum field theory of elementary particles. The numerical value of α defines how strongly photons and electrically charged particles interact through the electromagnetic force, which controls the universe along with gravity and the strong and weak nuclear forces. Among its effects, electromagnetism determines the degree of repulsion between protons and how electrons behave in an atom. If the value of α were much different from the one we know, that would affect whether nuclear fusion within stars produces the element carbon or whether atoms can form stable complex molecules. Both are necessary for life, another reason α is significant.

Other constants represent other major physical theories: c, the speed of light in vacuum, is crucial in relativity; h, the constant derived by Max Planck (now taken as "h-bar," or $\hbar = h/2\pi$), sets the tiny size of quantum effects; and G, the gravitational constant in Newton's theory and general relativity, determines how astronomical bodies interact. In 1899 Planck used just these three to define a universal measurement system based on natural properties and not on any human artifacts. This system, he wrote, would be the same "for all times and all civilizations, extraterrestrial and non-human ones."

Planck derived natural units of length, time, and mass from c, $\hbar$, and G: $L_P = 1.6 \times 10^{-35}$ meters, $T_P = 5.4 \times 10^{-44}$ seconds, and $M_P = 2.2 \times 10^{-8}$ kilograms. Too small to be practical, they have conceptual weight. In today's universe the gravitational interaction between elementary particles is too weak to affect their quantum behavior. But place the bodies a tiny Planck length L_P apart, less than the diameter of an elementary particle, and their gravitational interaction becomes strong enough to rival quantum effects. This defines the "Planck era" 10^{-44} seconds after the Big Bang, when gravitational and quantum effects were of similar strength and would require a combined theory of quantum gravity instead of the two separate theories we have today.

Nevertheless, to some physicists, c, $\hbar$, and G are not truly fundamental because they depend on units of measurement. Consider for instance that c is 299,792 km/sec in metric units but 186,282 miles/sec in English units. This shows that physical units are cultural constructs rather than inherent in nature (in 1999, NASA's Mars Climate Orbiter fatally crashed because two scientific teams forgot to check which measurement system the other had used). Constants that are pure numbers, however, would translate perfectly between cultures and even between us and aliens with unimaginably different units of measurement.

The fine structure constant α stands out as carrying this favored purity. In 1916 it appeared in calculations for the wavelengths of light emitted or absorbed as the single electron in hydrogen atoms jumps between quantum levels. Niels Bohr's early quantum theory predicted the main wavelengths but spectra showed additional features. To explain these, the German theorist Arnold Sommerfeld added relativity to the quantum theory of the hydrogen atom. His calculations depended on a quantity he called the fine structure constant. It includes $\hbar$, c, and the charge on the electron e, another constant of nature; and the permittivity ε_0 that represents the electrical properties of vacuum. Remarkably, the physical units in this odd collection cancel out, leaving only the pure number 0.0072973525693.

Sommerfeld used α just as a parameter, but it gained fame in the late 1920s when it reappeared in advanced work on relativistic quantum mechanics by the French physicist Paul Dirac, and then in what the English astronomer Arthur Eddington hoped would be a Theory of Everything. He planned to merge quantum theory and relativity to derive the properties of the universe such as the number of elementary particles in it, and its constants, α among them.

One twist in Eddington's approach was that he considered the quantity $1/\alpha$ rather than α, because his analysis showed that it must be an integer as well as a pure number. This was consistent with a contemporary measurement that yielded $1/\alpha$ = 137.1, tantalizingly near 137 exactly. Eddington's calculations gave instead 136, close enough to raise interest. Further measurements however confirmed that $1/\alpha$ = 137.036. Eddington's attempts to justify his different result were unconvincing and for this and other reasons his theory has not survived.

But α and "137" remain linked, which is why Richard Feynman called 137 a "magic number." What he meant has nothing to do with numerology. Rather it is that we know how to measure the value of α but not how to derive it from any theories we know. This is true also for the other fundamental constants, including pure numbers such as the ratio of the proton and electron masses, and is a lack in the Standard Model. Nevertheless, the value of α is critical in quantum electrodynamics, the quantum theory of electromagnetism. Feynman fully understood this, since he earned the 1965 Nobel Prize with two other theorists for developing quantum electrodynamics.

So α is accepted as one of the important constants of nature. Now, with the values of these quantities accurately known, physicists ask, are they truly constant? In 1937, considerations about the forces in the universe led Dirac to speculate that α and G change with time as the universe ages. Another suggestive and even older speculation is to wonder whether the constants vary across the universe. In 1543, when the Polish astronomer Nicolaus Copernicus put the Sun and not the Earth at the center of the universe, he moved humanity from its special cosmic location. This implies that the universe is the same everywhere, but this is only an assumption.

Varying "constants" would alter both the Standard Model and the cosmology based on it and general relativity, which among other issues fail to explain dark matter and dark energy. Add the role of α in the notion that the universe is "fine-tuned" to support life and the related idea that out of many multiverses, the one where we exist is the one with that winning value of α. All this spurs research on the constants of nature, much of it focused on α.

Earthly measurements confirm that α is fixed to within parts per tens of billions. A more challenging project is measuring it over astronomical distances. This also determines α at early cosmic times, since light from billions of light-years away took that many years to reach us from a younger universe. Since 1999, John Webb at the University of New South Wales, Australia, with colleagues has been making such measurements by gathering light from the distant galactic cores called quasars where black holes pull in dust that glows. This light traverses interstellar gas clouds and is absorbed at wavelengths characteristic of the atoms in the clouds. Analyzing the wavelengths gives α at the distant location, just as hydrogen wavelengths first defined α on Earth.

Webb's early results showed that α has increased 0.0006% over the last 6 billion years or more, and that it depended on distance from the Earth. Results published in 2020 show a smaller change in α between now and 13 billion years ago, when the universe was only 0.8 billion years old, which the authors interpret as "consistent with no temporal change." The cumulative results also suggest that α varies along different directions in space. Overall, the experimental errors are too large to inspire confidence that any single measured change in α is exactly correct, but the changes are certainly extremely small.

Now α has also been measured within a strong gravitational field, where it can theoretically change. The strongest gravity we know comes from a black hole, where a spacecraft would have to reach the unattainable speed of light to escape. But strong gravity also accompanies a white dwarf, a star that has expelled its outer layers to leave a massive but only planet-sized core. In 2013, J. C. Berengut of the University of New South Wales, with Webb and others analyzed spectral data from a white dwarf and obtained a change in α of 0.004% relative to the Earth.

No one, however, had measured α near a supermassive black hole until this year's work by Hees and co-authors including Ghez. Her results from Keck helped in choosing five stars whose orbits bring them near the supermassive black hole to maximize its gravitational effects, and of a type whose spectra display strong absorption features due to the surrounding stellar atmosphere. This facilitated deriving α from the absorption wavelengths for each star. The final composite result again shows only a small change in α, of 0.001% or less compared to Earth.

Although the measured change in α is small, the results for five stars at different places in the supermassive black hole gravitational field led to a new outcome; they allowed an early test of theoretical predictions that changes in α are proportional to changes in the gravitational potential, the energy stored in the gravitational field. The results confirmed that the two quantities are proportional, but uncertainties in the data supported only a rough estimate of the proportionality constant. A more reliable value could select between several new theories that treat dark matter and dark energy.

For now, the measured changes in α across time and space, and under gravity, are too small or uncertain to guide physicists toward new theories or even to fuel speculation such as the prospects for life far out in the universe or near a black hole. The smallness of the changes hints at the Copernican view that on very large scales, the universe looks much the same everywhere, although more measurements could confirm that there are real if tiny differences across the universe that might be meaningful.

It may be a kind of comfort to know that in a dynamic universe, this particular cosmic number holds steady. But seeing larger changes in α near our galactic supermassive black hole could be the starting point for new physics. As Hees described in an email interview, his goal now is to get deeper into the black hole's gravitational field. He plans to carry out new optimized measurements in 2021 to "observe stars that are closer to the black hole and therefore that experienced a stronger gravitational potential ... But with current technology, it is not easy to get good spectral observations of stars that are super close to the black hole." Still, he believes he can reduce measurement errors by a factor of 10.

The world-class Nobel Prize work carried out by Ghez depended on vast improvements in observational and spectroscopic techniques. It is a good bet that further improvements building on this successful project will enhance research in the supermassive black hole, a unique arena to study those elusive changes in α and what they mean for how we understand the universe.

The Most Popular Physics Meme Ever

The term "meme" comes from the Greek word *mimeme* ("that which is imitated"), and it denotes an entity (such as an idea or symbol) that propagates from mind to mind within a culture. The term was coined by Richard Dawkins in his bestselling book about evolution, *The Selfish Gene* (1976), and Dawkins originally saw memes as functioning like biological genes in their properties of self-replication, mutation, and so on. Not everyone accepts this analogy, but as a concept, the meme has nevertheless gained wide acceptance. It is routinely used, for instance, to describe the online process of "going viral."

But memes have meatier uses than merely explaining the replication of cute kitten photos. In a recent study of scientific memes [1], Tobias Kuhn of ETH Zurich and his colleagues used massive computing power to analyze close to 500,000 titles and abstracts from the *Physical Review* (spanning the period from 1893 to 2008) along with more than 46 million papers in Web of Science and PubMed. Their goal was to investigate the relationship between how often memes occur and how far they spread, and they began by finding the *n*-grams—strings of one to *n* words—that appeared most often in each database. To evaluate the propagation of the ideas that these represent, the authors then measured the appearance of each *n*-gram in papers that cited other papers that included the same *n*-gram. Then they multiplied this "propagation score" by the *n*-gram's frequency of occurrence to produce a "meme score."

After showing that meme scores behaved similarly in all the databases, the authors focused on physics, listing the 50 highest-scoring physics memes in descending order from "loop quantum cosmology" down to "Higgsless." One striking feature of the list is the way it mixes terms such as "black hole," which most physicists could identify and discuss, with narrowly focused ones like the chemical formula "MgB_2," the significance of which is best known within particular research areas (it is a superconductor). When I asked Kuhn about this, he replied that the analysis "finds relevant phrases not only on the global level but also for small sub-fields...the scale of distribution does not matter, only the distribution process."

This is an advantage, he believes, because terms that are used across many fields may be too common to be interesting as memes.

Meme analysis also examines how scientific ideas percolate into public understanding. A full study of "public memes" would be a major project, but the work by Kuhn's team provides clues: 38% of the 50 memes, they found, appear in physics articles in *Wikipedia*, and high-scoring memes correlate with article titles in the online encyclopedia. These appearances do not guarantee understandability of the concepts, since many *Wikipedia* articles are quite technical, but they do show that memes can reach highly popular outlets.

Another way to study the popularity of physics memes is through the online database Google Ngram, which contains 500 billion words in several languages, taken from digitally scanned books. When a user enters a word or phrase, Ngram plots its incidence as a percentage of the words in the database versus time for any period between 1500 and 2008, making it possible to examine the evolution of words as well as their prevalence.

I used Ngram a few years ago to study scientific and pseudo-scientific terms (October 2011 *Physics World* Lateral Thoughts), but this time I used it to test the penetration of Kuhn's 50 physics memes in the popular consciousness. Initially, I entered the memes just as they were written, but when I found some that did not appear or appeared only once, I also tried alternative forms of words or replaced low-scoring memes with less specialized terms that embody similar concepts. For instance, "loop quantum cosmology" does not appear in Ngram but "quantum cosmology" does. The result is that 22 of the 50 original memes, or their variants, have spread beyond the physics community to appear as physics-specific terms in Ngram. They aren't especially prevalent—in 2008, the most widespread Ngram physics meme comprised just 0.000024% of the database, compared with the 4.6% incidence of the word "the"—but they are there.

So what is that most popular meme? To the surprise of no-one who has seen *Interstellar*, the Ngram meme that apparently best grabs the popular fancy is "black hole." The field of runners-up, though, contains a few surprises. The second most common term is "nanotube," followed by "quantum dot," "molecular dynamics," "traffic flow," phrases containing "Higgs" (even before the discovery of the Higgs particle in 2012) and "dark energy." The Ngram memes are also quite varied, incorporating some topics that are theoretical or esoteric ("quantum entanglement"), but also some that are experimental or technological ("sonoluminescence," "graphene").

My admittedly limited survey shows that important physics ideas of any kind can enter the general consciousness. This is important at a time when science has increasingly shown its worth to society, and it could suggest new ways of presenting science as a public good. My analysis is also a step toward an idea suggested by Kuhn, who proposed establishing a monitor of trending *n*-grams to keep the public informed about scientific progress. Until such a tool exists, the Ngram physics memes offer guidance to science-fiction writers: if you include a spacecraft made of *graphene*, whose crew somehow use *nanotubes* and *quantum dots* to explore *dark energy*, *black holes*, and *quantum cosmology*, you can't go too far wrong.

Reference

1. T. Kuhn, M. Perc, and D. Gelbing, *Phys. Rev. X* **4**, 041036 (2014).

Can Space Experiments Solve the Puzzle of Quantum Gravity?

The biggest achievements of 20th and 21st century physics are quantum mechanics (QM), which led to the Standard Model that describes the behavior of elementary particles such as quarks, protons, and electrons; and general relativity (GR), Einstein's theory of gravity that explains the interactions between large cosmic bodies. Between them, the two theories cover all the significant processes of the Universe, which are determined by only four forces. Besides gravity, which keeps planets and comets in their orbits and operates on stars and galaxies as well as within black holes, there are the so-called strong and weak nuclear forces that rule quarks and atomic nuclei, and the electromagnetic force that controls atoms and molecules.

But there is a deep problem: the Standard Model explains the last three forces, but not gravity, which is separately explained by GR. Although each approach works well in its own regime, physicists try to make theories as universal as possible, and so theoreticians seek a master theory of quantum gravity that explains all four forces from a single viewpoint. But merging the two theories has so far been unsuccessful. A breakthrough however may come as researchers begin carrying out quantum experiments in space that give insight into how QM and GR interact, but that cannot be done on Earth.

The merger was never going to be easy because the two theories embody utterly different approaches. The three forces in the Standard Model arise from the exchange of certain "force-carrier" particles, like a crowd of children pelting each other with snowballs. The electromagnetic force comes from the exchange of photons, the quantum particles of light, and the strong and weak forces similarly arise from the interchange of other specific particles. But in GR, Einstein's insight was to realize that gravity comes from the structure of spacetime. In empty space, spacetime is undisturbed and the natural path of an object is a straight line. But a large mass like a star distorts spacetime so that the natural path of an object becomes curved, such as an orbit around the star. The theory of gravity depends on the geometry of spacetime, not on force-carrier particles.

However, among attempts to reconcile the two different views, string theory has appeared promising in principle. Its premise is that elementary particles are tiny vibrating strings rather than mathematical points, with different vibrational modes representing different elementary particles. Besides the known particles, the mathematics of string theory predicts a new one, dubbed the graviton, the carrier particle for gravity. This would allow a seamless merger of QM and GR into a theory of quantum gravity. But there are no known experiments that can confirm string theory in general or the existence of the graviton, as would be necessary to turn purely mathematical results into physical reality, nor have other approaches yielded a satisfactory theory of quantum gravity.

Now however the availability of platforms for scientific investigations in space makes it possible to carry out quantum experiments under varying gravity to give clues to quantum gravity. One platform is ISS, the International Space Station, operated by five national space agencies including NASA and the European Space Agency ESA. Orbiting the Earth at an average altitude of 400 kilometers, it is inhabited long-term by astronauts who carry out varied experiments. Another platform is Micius, a satellite in orbit 500 kilometers above the Earth. Micius was established by the Chinese Academy of Sciences to develop quantum science and technology in space, for goals such as designing a system of Earth satellites for secure global communications.

The basic quantum gravity experiment uses entangled pairs of photons. Entanglement is a quantum effect where two photons, once created as a pair, remain forever linked even if far apart. For one example of the effect, consider the electric field of a photon, which can point either horizontally (H) or vertically (V). An entangled photon pair has one H and one V photon but we do not know which is which. However, whichever value results when measuring photon A, photon B takes the other value. This is as if you and a friend who is miles away each had a dresser drawer containing black and white socks in equal numbers. If the socks were entangled, whatever color sock you randomly pick from your drawer, your friend would randomly pick the other color every time, defying the conventional odds. The two socks—or two photons—are correlated in ways that do not arise in classical physics.

Another form of entanglement is more suitable for space experiments. In energy–time entanglement, measuring the frequency of photon A (which is proportional to its energy) at one point in space determines when its partner photon B arrives at a different location. This kind of entanglement was used in the first space quantum gravity experiment, carried out in 2019.

A team under Jin-Wei Pan of the University of Science and Technology of China prepared entangled photon pairs at a ground station in Tibet. One photon from each pair was kept there, while its entangled partner was sent up to Micius where automated equipment registered its arrival time. This was compared to the frequency of its partner as measured on Earth to see if the pairs were correlated as predicted by energy–time entanglement. In standard QM, whether the two photons in each pair experienced the same or different gravitational fields, they would remain entangled. But according to the theory of event formalism (developed by Timothy Ralph of the University of Queensland, Australia, and Jacques Pienaar of the Austrian Academy of Sciences), when paired entangled photons traverse different gravitational fields, a certain fraction of the pairs will decohere, that is, lose their entanglement.

The results from the Micius experiment showed no decoherence within the experimental errors, which means that standard QM need not be altered to include gravitational effects, that is, to accommodate GR. This is one useful clue in the search for a theory of quantum gravity, but only if event formalism theory is correct. Moreover the research team notes that because of some technical limitations aboard the satellite and assumptions made in interpreting the data, they plan further measurements to obtain a completely definitive result. These new efforts will likely benefit from the Micius team's ongoing work toward developing secure quantum communications, such as new results presented in 2020.

The Micius team however is not the only group to pursue these studies. In 2018, Siddarth Joshi of the Austrian Academy of Sciences and 44 international colleagues proposed a project they called QUEST (Quantum Entanglement Space Test). This would carry out space entanglement measurements aboard ISS, where the presence of a crew that can assist with the experiment could provide useful flexibility. The project requires approval from ESA, which the team is now awaiting. In an email interview Joshi told me that should this be forthcoming, the first step would be to replicate the Micius experiment. He believes that knowledge gained from that experience would allow QUEST to carry out an improved effort that could conclusively confirm the Micius result, or yield new information. In any case, two simultaneous approaches to these difficult experiments represents significant research momentum.

In a broad view of the value of space science and technology as outgrowths of space exploration, these efforts to shed light on quantum gravity add to a growing list of varied and sometimes unexpected benefits such as improved astronomical observation, and closer to home, quantum communication and insights into human health and longevity. If space experiments were also to provide important clues to unifying quantum mechanics and general relativity, space science could chalk up a major contribution to solving a long-standing scientific riddle.

Small Wonders

Review of *Taming the Atom: The Emergence of the Visible Microworld*, Hans Christian von Baeyer (Dover Publications, 2000)

Although nature enfolds the huge and the minute, and both extremes grew from the Big Bang, human reactions to the two could not be more different. Size creates awe, even primal dread. It is the need to find a niche amidst vast echoing spaces that drives us to understand the universe in the large. But we feel no cosmic dread of the small. Our desire to examine it comes instead from the ancient idea that to know the building blocks is to know all.

The path to the small is blocked by perceptual barriers that do not exist on the road to the large. One barrier is that the ordinary light that lets us examine clusters of galaxies is of too great a wavelength to discern anything so fine as an atom. And the quantum theory that describes the microcosm has its own perceptual pitfalls. Its seemingly confused melding of wave and particle is only the beginning. If we do stand in awe of the small, it is because we see its complexities through so dim a glass. In *Taming the Atom*, Hans Christian von Baeyer clarifies our vision. A theoretical physicist who shares his understanding in graceful prose, he guides us into the realm of atoms, the smallest bits of matter we can truly examine. Contrary to what we used to learn in grade school, it is now possible to "see" and "feel" individual atoms. This leads to the book's central message, or at least its main hope, that the new fineness of perception will help resolve the contradictions in quantum mechanics that have puzzled physicists from Einstein on.

One new way of seeing described in *Taming the Atom* is the Nobel Prize–winning technique of scanning tunneling microscopy, in which a tiny metal stylus emits electrons to map the contours of a solid surface a few atomic diameters away. Although the resulting portraits of atoms provide an undeniable thrill, they do not yet offer enough detail to say much beyond "the atom is located there." Even when the resolution of our micro-cameras improves, exploring inner atomic workings will require that detailed pictures be combined with other experiments and mathematical analysis.

Nevertheless, von Baeyer believes that it is through these avenues that physicists will eventually come to understand such quantum enigmas as how electrons can behave both as waves and as particles. To illustrate this property, imagine that a beam of electrons encounters a barrier containing two parallel slits. If electrons were simply particles, some would flash through each slit in straight lines, like tiny bullets. The two streams exiting the slits would never meet; and even if they did, they could not cancel each other out. But in fact the streams behave like waves, fanning out to meet and canceling each other's undulations at certain points in space. The resulting "interference pattern" appears as a set of alternating bright and dark lines on a fluorescent screen that intercepts the electrons, glowing where they hit, remaining dark where they do not. Yet at the same time, a detector at the screen emits discrete "pings" as if the electrons were arriving bullet by bullet.

This schizoid behavior is not good for physicists' peace of mind, and leads to even deeper unrest. Consider this scenario, which the author treats at length: Suppose the electrons come at such a leisurely pace that only one at a time encounters the twin slits, and many minutes pass before the next one arrives. As hours go by, the expected bright-dark striping of wave interference slowly forms on the screen. Why? Through some mysterious process, an electron acts as if it "knows" that it should avoid certain spots on the screen—thereby setting the stage for an interference pattern— long before its fellows arrive. No wonder the Nobel physicist Richard Feynman despaired that "nobody understands quantum mechanics." Only a brand-new concept, it seems, can free us from the tyranny of separate wave and particle. But although the author discusses experiments, especially in the area of superconductivity, that may yield this breakthrough idea, it remains hidden.

One promising avenue that von Baeyer does not tread in his search for the theory of the small is the study of technologies that are actually based on quantum mechanics. Experimental physicists like me find that quantum theory is woven into our daily scientific and technical life. The first time I tuned a carbon-dioxide laser, watching its intense light disappear at one wavelength and reappear at another, I felt a shock as I saw the reality of quantized energy levels. Even the lowly light-emitting diode, a scrap of semiconductor cheap enough to decorate nearly every piece of electronic equipment made, applies the mysterious quantum principles. Its electrons act as both particle and wave to generate another strange amalgam, the combined wave and particle of light.

But it is the cutting edge that is most likely to hold the solutions to quantum puzzles. We can now construct matter virtually atom by atom to make better superconductors for efficient use of electricity or artificial semiconductors for faster electronic devices. These custom-made materials become micro-laboratories where the quantum world can be more readily examined – to answer old riddles, but also to produce materials that may be more useful still. In times when pure science is bluntly asked to give an accounting of itself, it is bracing to think that we can improve technology for society while investigating the roots of the microcosmic.

Where does this leave the other end of the scale? In the long view, there is no distinction: the quantum must somewhere join the galaxy, perhaps in a theory of quantum gravity, perhaps at the core of a black hole. But the two extremes are already linked in a satisfying way, because we use the small to seek the large. Telescopes, our cosmological probes, "see" with electronic sensors that gently coddle the few photons arriving from distant sources. These sophisticated cousins of the light-emitting diode have reached their remarkable sensitivity through applied quantum mechanics and the use of techniques like scanning tunneling microscopy. Can anyone doubt that it is all connected?

The Quantum Random Number Generator

Maybe this has never crossed your mind, but if you have ever tossed dice, whether in a board game or at the gambling table, you have created random numbers—a string of numbers each of which cannot be predicted from the preceding ones. People have been making random numbers in this way for millennia. Early Greeks and Romans played games of chance by tossing the heel bone of a sheep or other animal and seeing which of its four straight sides landed uppermost [1]. Heel bones evolved into the familiar cube-shaped dice with pips that still provide random numbers for gaming and gambling today.

But now we also have more sophisticated random number generators, the latest of which required a lab full of laser equipment at the U. S. National Institute of Standards and Technology (NIST) in Boulder, CO. It relies on counterintuitive quantum behavior with an assist from relativity theory to make random numbers. This was a notable feat because the NIST team's numbers were absolutely guaranteed to be random, a result never before achieved.

Even so, you might ask why random numbers are worth so much effort. As the journalist Brian Hayes writes in "Randomness as a Resource," these numbers may seem no more than "a close relative of chaos" that is already "all too abundant and ever present" [2]. But random numbers are chaotic for a good cause. They are eminently useful, and not only in gambling. Since random digits appear with equal probabilities, like heads and tails in a coin toss, they guarantee fair outcomes in lotteries, such as those to buy high-value government bonds in the United Kingdom. Precisely because they are unpredictable, they provide enhanced security for the internet and for encrypted messages. And in a nod to their gambling roots, random numbers are essential for the picturesquely named "Monte Carlo" method that can solve otherwise intractable scientific problems.

When you log into a secure internet site like the one your bank maintains, your computer sends a unique code to identify you to the responding server. If that identifier were the same every time you logged on, or were an obvious sequence like 2468 or QRST, a hacker might retrieve it or deduce it, impersonate you online, and take your money. But if the identifier is a random string that is newly created for each user and each online session, it becomes impossible to hack.

This idea shows up in a two-step verification, when after presenting a site with your password, your phone receives a random multi-digit number that you must enter to complete the login. The same logic applies for an encrypted message. The person who receives it needs the key in order to read the message. This key must be transmitted from sender to receiver, and so is vulnerable to hacking. But if the key for *each* message is a *new* random string, that protects messages from being compromised through knowledge of their keys.

However, making random numbers for these or any other purposes is not easy. Supposedly random strings invariably turn out to be flawed because they display patterns. For instance, in 1894, the English statistician W. F. R. Weldon, a founder of biostatistics, tossed dice more than 26,000 times to test statistical theory [2]. And in 1901, the great English physicist Lord Kelvin tried to generate random digits by drawing numbered slips from a bowl. But Kelvin could not mix his slips well enough to make each digit appear with equal probability. Later analysis showed that Weldon's dice were unbalanced, yielding too many fives and sixes compared to the other numbers.

World War II provided another example of imperfect randomness with real-world consequences. It came from the famous effort at Bletchley Park in England, where cryptographers and mathematicians such as Alan Turing worked to break the German military code. The Germans used the typewriter-like Enigma machine to encode messages [3], replacing each letter of the alphabet with another letter in a way that changed for each letter in the message, accomplished by a set of rotating wheels that clicked into one of 26 positions labelled A to Z. Before typing a message to be encoded, the operator would randomly choose initial positions for the wheels and send that in code as the key to the coming message. The receiving operator set his Enigma to the same key to correctly decode the message as each incoming letter was typed.

But the operators often became careless or took shortcuts, using the same key for several messages or selecting initial letters near each other on the keyboard. These breaks from randomness, with other clues, helped the Bletchley Park team analyze intercepted Enigma messages and finally break the code, contributing to the Allied effort against the Nazis.

Advanced random number generators were developed later. In 1955, the Rand Corporation published a million random digits from an electronic circuit that, like spins of a virtual roulette wheel, created supposedly unpredictable numbers. But these too displayed subtle patterns. Another approach accompanied the beginnings of digital computation in 1945 with the first general-purpose electronic computer ENIAC (Electronic Numerical Integrator and Computer) at the University of Pennsylvania.

Computerized randomness came from some of the best scientific minds of the era, associated with the Los Alamos Laboratory of the Manhattan Project to build an atomic bomb: Enrico Fermi, a Nobel Laureate for his work in nuclear physics; John von Neumann, considered the leading mathematician of the time; and Stanislaw Ulam, another mathematician who, along with Edward Teller, invented the hydrogen bomb.

In the 1930s, Fermi had realized [4] that certain problems in nuclear physics could be attacked by statistical means rather than by solving extremely difficult equations. Suppose you want to understand how neutrons travel through a sphere of fissionable material like uranium 235. If enough neutrons encounter uranium nuclei and split them to release more neutrons and energy, the result can be a chain reaction that runs away to become an atomic bomb— or that can be controlled to yield manageable power.

Fermi saw that one approach to the problem was to guide an imaginary neutron along its path according to the odds that it would encounter a uranium nucleus. It would then be deflected in a new direction or absorbed by the nucleus to generate more neutrons. If the calculation were done for each of a horde of primary neutrons, the secondary neutrons they produced, and so on, using random numbers to select the initial neutron velocities and the options they encountered as they traveled, the result was a valid picture of the whole process.

In 1947, von Neumann and Ulam began simulating neutron behavior on ENIAC. Von Neumann produced random numbers with a computer algorithm that began with an arbitrary "seed" number and looped to produce successive unpredictable numbers. This "Monte

Carlo" approach (a name supposedly suggested by the fact that an uncle of Ulam's used to borrow money to gamble there) proved successful for the neutron problem [4]. The method became widely applied, and it established the scientific validity of computer simulations for certain types of problems [5]. "Computational physics" is now an important tool for physicists, and Monte Carlo methods contribute in areas as diverse as economic analysis and ornithology [6, 7].

But as von Neumann knew, the numbers from his algorithm were not really random. They repeated after many iterations, and the same seed always created the same numbers. Von Neumann commented, tongue-in-cheek but also to make a point, that anyone using this procedure, himself included, was in a "state of sin" [8]. He meant that the procedure violates a philosophical truth: a deterministic process, such as a computer program, can never produce an indeterminate outcome. "Random" numbers made by computer are valid in some cases if used with care, but they are better called "pseudo-random." And it is pseudo-random numbers that appear in security applications. Since hackers could uncover the seed number and the algorithm that uses it, these numbers are predictable and do not give absolute security.

The random numbers made at NIST's Boulder labs in 2018, however, are not "pseudo" because they come from the inherent indeterminacy of the quantum world. The scientist leading the project, Peter Bierhorst (now at the University of New Orleans), made these numbers by applying the quantum effect called entanglement to photons [9]. A photon is a quantum unit of electromagnetism that carries an electric field. This can point either horizontally or vertically, representing a computer bit with value 1 or 0. Since quantum mechanics is statistical in nature, either direction of the field, called the polarization, or either value 1 or 0, has the same 50% probability of appearing when measured.

When two photons are entangled, their states of polarization are linked and are opposite. If the polarization of one photon is measured, whatever the result, the other one, no matter how far away, takes on the opposite value. There is no physical connection between the photons. Strange as entanglement seems, it is a real, purely quantum correlation that is independent of distance.

In Bierhorst's scheme, a laser created pairs of entangled photons. Each photon in a pair went to one of two separate detectors that recorded its polarization as a 1 or a 0. These detectors were placed 187 meters (nearly two football fields) apart. The fastest possible signal between them, a radio wave moving at the speed of light, would need 0.62 microseconds to cover the distance. This was more than the time it took a detector to register 1 or 0, ensuring that the two sets of measured bits could not be affecting each other by any conventional means [10].

Nevertheless, the photons were correlated because they were quantum entangled, as the researchers confirmed. Since the bits read from photon A exactly track the bits read from photon B over any distance whatsoever, this suggests the possibility of faster-than-light communication. But relativity forbids such communication, so it must be that the photons arrive at a detector with 1s and 0s in random order, given that each possible value occurs exactly half the time.

The result was the first set of measured binary digits that can be unequivocally certified as random. These unpredictable numbers would supply total security and completely valid statistics if they replaced pseudo-random numbers in applications. NIST plans to integrate the quantum procedure into a random number generator that it maintains for public use, to make it more secure. Meanwhile, Bierhorst is working to reduce the physical size the arrangement needs to guarantee that there is no conventional communication.

Apart from its technological value, this quantum random number generator would have intrigued Einstein. He would probably be pleased that relativity theory contributed to the result; but the process also says something about quantum theory, whose random nature Einstein rejected when he said that God does not play dice with the universe. Maybe not, but humanity has worked out how to make its own superior, quantum-based, and truly random dice.

References

These references can be found at their individual sources or in the JSTOR digital archive at https://www.jstor.org/.

1. D. R. Green, Historical topics: The beginnings of probability theory, *Mathematics in School*, Vol. 10, No. 1 (Jan. 1981), pp. 6–8.

2. B. Hayes, Computing science: Randomness as a resource, *American Scientist*, Vol. 89, No. 4 (Jul.–Aug. 2001), pp. 300–304.

3. F. Churchhouse, *Codes and Ciphers: Julius Caesar, the Enigma, and the Internet* (Cambridge University Press, Cambridge, 2001).

4. N. Metropolis, The beginning of the Monte Carlo method, *Los Alamos Science*, Special Issue, 1987, 125–130. https://library.lanl.gov/cgi-bin/getfile?00326866.pdf.

5. D. D. McCracken, The Monte Carlo method, *Scientific American*, Vol. 192, No. 5 (May 1955), pp. 90–97.

6. E. Taylor et al., Economic impact of refugees, *Proceedings of the National Academy of Sciences of the United States of America*, Vol. 113, No. 27 (July 5, 2016), pp. 7449–7453.

7. A. L. Hughes, J. Cauthen, and C. Driscoll, Testing for behavioral lateralization in observational data: A Monte Carlo approach applied to neck-looping in American flamingos, *The Wilson Journal of Ornithology*, Vol. 126, No. 2 (June 2014), pp. 345–352.

8. G. E. Forsythe, Summary of "various techniques used in connection with random digits" by John von Neumann, *Journal of Research of the National Bureau of Standards*, Series 3 (1951), pp. 36–38.

9. L. Ost, NIST's new quantum method generates *really* random numbers, *NIST News*, Apr. 11, 2018. https://www.nist.gov/news-events/news/2018/04/nists-new-quantum-method-generates-really-random-numbers

10. P. Bierhorst et al., Experimentally generated randomness certified by the impossibility of superluminal signals, *Nature*, Vol. 556 (2018), pp. 223–226.

Sunlight, Life, and Time

Reviews of the films *Brilliant Noise, Solar Sinter Project, Photo Synthesis*, and *Stuck in the Past* (these films can be seen at the Labocine site https://www.labocine.com/).

Film and video are the arts of putting meaningful dynamic images on screen, images that could not be created and appreciated without the visible light we humans see. It's fitting that some Labocine films use the art of moving light to tell us about light itself and its greatest source for us, the Sun, which dominates earthly life—and by the way, made movie making possible.

In 1893, Thomas Edison opened the first film studio, located in New Jersey and sporting a black exterior that earned it the name Black Maria (slang for a police van). The carbon arc lamp and Edison's own invention, the incandescent light bulb, existed then but the first was too erratic and the second too weak to properly expose the insensitive film stock of the time. Instead the Black Maria had a large retractable roof to let in sunlight and was mounted on a turntable so it could follow the Sun all day. In the same vein, when the movie business moved west to Hollywood in the early 20th century, it was partly because California offered nearly constant sunshine and a pleasant climate in which to shoot movies in natural light.

Sunshine played a benign role in the birth of film but it has a dark side in causing skin cancer, reflecting its violent origins. The Sun is a hellhole of heat and activity, a ball of plasma 1.4 million kilometers across where nuclear processes produce energy that keeps its interior at 15 million Kelvin. Nothing human could survive for a nanosecond in its maelstrom of particles and radiation. Even the Sun's surface, at nearly 6000 Kelvin, is too hot for human-made structures. Along with its visible glow, the surface radiates the ultraviolet light that causes cancer, other wavelengths with various effects, and particles that reach the Earth as the solar wind.

Brilliant Noise (Semiconductor, 2008, U. K., 6 minutes) dramatically shows the Sun's violent nature. Combining snapshots from observatories at Mt. Wilson, California, and elsewhere, it shows striking black and white time-lapse movies, from streams of charged particles leaving the solar sphere to close-ups of loops of light bigger than the Earth emitted from particles that follow magnetic lines of force near the surface. These are associated with the solar wind and with sunspots, the darker areas that appear for reasons not understood and produce solar storms that can disrupt radio communication on Earth. An audio track enhances the images by converting their changes in visual intensity to sounds and radio static. The result is a powerful representation of the tumult in and on the Sun.

That turmoil is tamed by the time sunlight reaches us, but still it carries enough energy to affect solid materials as well as our own skin, as shown in *Solar Sinter* (Markus Kayser, 2011, U. K., 6 minutes). The film opens with a shot of nothing but sand out to the horizon in the Sahara Desert. Soon a young man dressed in white—Kayser himself—drags an unwieldy but interesting-looking structure into the frame. It supports solar cells, a battery, sensors, and electronics that track the Sun during the day, and a Fresnel lens—the same focusing device used in lighthouses to project a concentrated beam of light.

Working alone in the barren immensity, Kayser focuses sunlight into an almost blindingly intense spot onto a bed of sand scooped up from the surrounding desert. With no audio track except the sounds of the desert wind and the solar tracking machinery, close-up shots show that the solar spot is hot enough to sinter the grains of sand, that is, melt them enough to make them coalesce into a solid mass without full liquefaction. Under computer control that amounts to 3D printing, Kayser produces first an intricate twisted form and then a bowl that is 15 centimeters across, composed of a kind of impure opaque glass made from silica (silicon dioxide, SiO_2), a main component of sand.

Kayser's film is meant to show the potential of 3D printing using freely available solar energy and raw material; but the bowl he made has been put on display in the Museum of Modern Art (MOMA) in New York City, and so the film is also a statement about the fruitful combination of technology, design, and art. And it shows that after traveling 150 million kilometers (93 million miles) from its inconceivable origins inside the Sun, sunlight still carries energy that can be applied in new ways to benefit humanity.

The most important benefit of sunlight is not artificial but natural and life-giving. That is the photosynthetic process, where life is "woven out of air by light," meaning that sunlight drives the reactions in plants that support the chain of earthly life. *Photo Synthesis* (Barry J. Gibb, 2006, U. K., 4 minutes) shows how. The film begins with a voice-over:

> What amazes me is that…this light travels all the way from the Sun through 93 million miles of space only to find Earth [with the energy] to cause the plants to grow, the trees to grow, to reach back, back into the sky.

The rest of the film shows photosynthesis in action beginning with NASA images of the active sun producing the light that will reach Earth. Then a different educational film-like narrator repeatedly intones the importance of "sunlight, air, water, and chlorophyll" to change carbon dioxide into the oxygen we breathe (and not mentioned, the carbohydrates that feed plants and eventually, us). Images and words describe the process from the microscopic level to actual growing green leaves, plants, and trees. This is set against an audio track like a beating heart to remind us that solar energy and photosynthesis are truly matters of life and death.

These Sun-driven processes offer a challenge to the filmmaker who would like to portray the Sun's effects on Earth in real time. That can't be done, because our view of whatever happens on the Sun, even if it were suddenly to go supernova, is delayed by the time it takes its light to reach us. Even at light's enormous speed of 300,000 kilometers per second, its trip from Sun to Earth takes eight minutes.

Stuck in the Past (Javier Diez and Moiya McTier, USA, 3 minutes, 2016) explains how every astronomical observation, not just studying the Sun, is a window into the past. McTier, a PhD astronomy student at Columbia University, introduces us to the photon, the basic particle of light. She shows how photons cover earthly distances so quickly that light seems to travel instantaneously—but not for the vaster distances of the universe. Standing on a balcony, she imagines that Manhattan as seen to the south is scaled up by many billions to astronomical size. Then looking downtown to Times Square, instead of seeing current activity, we might see news of the sinking of the *Titanic* in 1912 just reaching us. Looking further south and so further back, the Brooklyn Bridge would not exist and New York would still be called New Amsterdam.

Similarly, astronomers observing stars and galaxies see not what they are now, but what they were up to billions of years ago depending on their distance. As this film's clear and engaging analogy shows, astronomers are "stuck in the past." So are all of us in terms of what we perceive, though on Earth we go back only tiny fractions of a second. But the telescopes that examine distant objects in the universe are time machines that can bring us nearer the actual instant of creation, a moment in the past we would all like to see.

These films show sunlight's special importance to us, but in the internal solar processes that make that light and in the time lag in observing the Sun, it also represents the workings of all active stars and the general laws of the universe. One lesson from these films is that while we and our own personal star may seem unique, the Sun and maybe life itself are simply parts of the universe that are found in many places and times.

Flash!

Lightning flashes through the sky millions of times a day, and humans have witnessed the spectacle repeatedly for tens of thousands of years. It was probably the spark for organic life, an idea tested in the lab back in 1952 when the chemist Harold Urey and his graduate student Stanley Miller exposed a simulation of the early atmosphere to artificial lightning. To their delight, they ended up with a primordial soup of amino acids—the building blocks of life. According to the sociobiologist Edward O Wilson, lightning even played a role in the evolution of the human mind. Wilson argues that pre-humans gained access to high-protein, brain-boosting meals when lightning burned big game on the African savanna, leaving entire carcasses cooked and ready to eat. That same lightning provided the flames that our ancestors captured to keep the home fires burning. It was around the campfire (enabled by lightning) that human philosophy and culture was spawned.

Early humans who saw only lightning's immense power, with no inkling of its electrical nature or its influences on life and evolution, believed that it came direct from the gods—most famously, the chief Greek god Zeus, who hurled down lethal thunderbolts from Mount Olympus. Norse mythology had Thor, son of Odin, whose rumbling chariot wheels and magic hammer generated thunder and lightning. Other cultures, from the Japanese to the Slavic, have honored similar deities.

These early associations might explain why lightning symbolizes the awe-inspiring power of nature and the creation of life. We are still gripped by the scene in the classic film *Frankenstein* (1931), where intense lightning flashes amid the roar of thunder animate the monster (Boris Karloff) made of dead body parts. Lightning also represents overwhelming force delivered at blinding speed. Military insignia often feature lightning bolts, and heavily armed military aircraft have been named Lightning and Thunderbolt since the Second World War. In that war, the Nazis called their fast-moving attacks that overran Europe *blitzkrieg*, for "lightning warfare"; they took the symbolism still further by stylizing the initials of their SS units as two lightning bolts.

"High voltage" signs still display fearsome lightning bolts but lightning has also become the universal symbol of electricity as our benign servant. A digital device such as my smartphone uses only small, harmless voltages, but the phone displays a tiny lightning bolt while it is being charged. For decades, the electric power industry advertised itself with Reddy Kilowatt, a stick figure with limbs made of lightning and a lightbulb for a nose.

The widespread electrical phenomenon of lightning has existed for much of the Earth's lifetime, but its origin story remains a mystery to this day. "When we consider how much we know about complex and exotic astrophysical objects half way across the Universe, it is quite amazing that we do not understand the basics of how something as common as lightning gets started in clouds just a few miles above our heads," wrote the physicists Joseph Dwyer and Martin Uman in a substantive review of the field in 2014. The irony is deepened by our urgent need to understand lightning, which has become more destructive around the globe, largely due to human activities and climate change.

The effort to see lightning through a scientific lens can be traced back to 6th century BCE and the Greek philosopher Anaximander, who attributed lightning and thunder to natural causes: fire, gusts of air, and collisions between clouds. Around 340 BCE, Aristotle similarly asserted that a "dry exhalation"—a combustible emission that arose from the Earth—caused thunder and then burned to produce lightning; we see the lightning first, he wrote, because "sight is quicker than hearing."

The mechanics were revealed only after 18th-century scientists began experimenting with electric sparks; the work helped them see that lightning was the same phenomenon—an electric spark between a cloud and the ground on an immense scale. The American polymath Benjamin Franklin was the first to suggest that this could be experimentally tested. He carried out the most famous such test in Philadelphia in 1752, when he launched a kite into a thunderstorm and held the end of the string where a metal key was attached. When he brought the knuckle of his hand near the key, he saw and felt an electric spark. That result began the modern era of lightning research.

Though work continued after Franklin's breakthrough, two centuries passed before the next major insight into lightning. In 1960, the German–American physicist Heinz-Wolfram Kasemir proposed that lightning begins when leaders—electrically active channels in the air where current can flow—develop between regions of positive and negative electric charge in a cloud. Unlike the simple, seemingly well-established idea of a spark from a charged cloud to the ground, Kasemir's idea meant that a complex series of steps precede what we see as lightning. His proposal was dismissed until measurements aboard research aircraft in the 1980s confirmed that there are, indeed, areas of positive and negative charge within and near thunderclouds.

Further knowledge came when researchers found ways to determine the voltage of a lightning bolt and the current that the voltage drives, the same properties that define any home electrical system. But at many millions of volts and thousands of amperes respectively, these quantities in lightning are orders of magnitude stronger than the 110 volts and 100 to 200 amperes that power the lamps, refrigerators, and TV sets in our homes.

Physicists learned to dissect the development of a bolt with high-speed cameras. Researchers went on to trigger lightning artificially by sending rockets trailing copper wire into thunderstorms. Dwyer, Uman, and other scientists at the International Center for Lightning Research and Testing in Florida (the U. S. state with the highest incidence of thunderstorms) have examined some 400 triggered lightning bolts over the years.

Through such efforts, researchers learned that lightning is produced when ice crystals in a thundercloud are carried upward in a rising air current to encounter particles of graupel, a kind of soft hail, along with super-cooled water that remains liquid below zero degrees Celsius. The interaction produces positive and negative charges respectively at the top and bottom of the cloud, and a positive region below it. The opposite charges grow and mutually attract until the electric field between them starts to break down the intervening air and a stream of electrons, an electric current, flows. This is the leader that Kasemir proposed, and it can occur within a cloud, between clouds, or from a cloud toward the ground.

High-speed cameras show that the most prevalent kind of lightning strike starts when a series of leaders (generally too dim to be seen by the naked eye) carries electrons downward in a zig-zag path. Nearing the Earth, the electrons interact with positive charges below, making the air fully capable of carrying current. As a result, electrons from the cloud race to the ground along the same jagged conduit that the leaders traced. This is what we perceive as lightning. The electron flow heats the air in its path to 30,000°C, five times the Sun's temperature, making a brilliant flash and rapidly expanding the air to create thunder. Other strokes can follow, but it is always the massive current of thousands of amperes that kills or injures living things and carries enough energy to burn or destroy what it hits.

Though we can trace the development of lightning in broad outline, there are some glaring lacks. Exactly how ice, water, and graupel interact to separate positive from negative charges is unknown. Another mystery: from all we know about electrical discharges, an intense electric field of millions of volts across the air gaps between the positive and negative charges is necessary before leaders can develop. Measurements within clouds indicate that the internal fields never reach this level—yet lightning happens. This is why Dwyer and Uman consider the birth of lightning among the biggest mysteries in atmospheric science.

Solving these puzzles would have practical and human meaning, because lightning is becoming more destructive. Franklin's invention of the lightning rod, which diverts the electric current of a bolt away from a structure, mitigated some dangers. But the potential of lightning to kill people and destroy structures, disrupt aviation and electrical systems, and ignite wildfires and forest fires has grown in the past decades. Much of this is due to human-induced global warming and atmospheric pollution. These encourage convection, the rising of warm air and moisture that feeds thunderstorms, and so produce more lightning. The atmospheric scientist Yoav Yair at the Interdisciplinary Center Herzliya in Israel points to another factor: the growing global concentration of people in urban areas—now at 55%, projected to reach 68% by 2050—where tall buildings and air pollution make lightning strikes more likely.

These changes can affect large areas. For example, Yair notes that disruption of air travel by lightning, in-flight and at airports generally located near big cities is a growing issue in East, South, and Southeast Asia. This Asia-Pacific region contains some major lightning-prone areas, while its air traffic is rising at nearly 5% annually. In another example, Australian and U. S. researchers analyzed lightning-induced fires across dry and wet ecosystems in Australia, South Africa, and South America. The number of such fires has grown in this century, a trend that scientists connect to climate change as it increases the number of lightning strikes and the resulting fires.

One response to these dangers that also serves research is to track lightning. A lightning flash is a powerful source of electromagnetic radiation, much of it at frequencies below 500 kilohertz (the AM radio band begins nearby at 540 kilohertz, which is why lightning produces static at the low end of the AM dial). Sensors operating in this range can quickly triangulate the location of a distant lightning strike. The World Wide Lightning Location Network (WWLLN), for instance, uses more than 70 devices placed internationally to trace lightning.

In 2017, Joel Thornton of the University of Washington and colleagues used WWLLN data to show that lightning occurs twice as often above two busy shipping lanes in the Indian Ocean and South China Sea as above adjacent lightly travelled ocean areas. The difference, the researchers surmise, comes from the aerosol particles that ships emit as they burn fossil fuel. This is strong evidence that atmospheric pollution enhances lightning. Lightning location systems also monitor extreme weather, protecting flight operations at airports. In 2013, a locating system tracked a powerful typhoon hundreds of kilometers distant from the island of Mindanao in the Philippines, where meteorological data were lacking.

Lightning produces other characteristic emissions in infrared wavelengths when it interacts with atmospheric oxygen and nitrogen. Invisible to the human eye, this radiation can be sensed from specially equipped NASA space satellites that scan large parts of our planet. In 2016, Rachel Albrecht of the University of São Paulo in Brazil and her colleagues conducted an extensive analysis of this satellite data. They identified areas in Africa and Asia as highly active lightning sites, and Lake Maracaibo in Venezuela as the most active area on Earth. Its climate and terrain generate thunderstorms 297 days per year, on average. These produce an annual average of 233 lightning flashes per square kilometer, compared with a global average of six flashes per square kilometer. The eventual goal is to combine such data with models of lightning activity for long-range predictions of where lightning will strike.

To protect against lightning, we also need to know the energy or power in a typical bolt. The lightning flash and associated thunder, the brightest light and loudest sound commonly experienced on Earth, indicate a high energy level. Another clue is that lightning creates X-rays and gamma rays. These are thought to arise because, like the Large Hadron Collider at CERN near Geneva, the electric fields accompanying lightning accelerate elementary particles— in this case, electrons—to high energies. These particles generate X-rays, and gamma rays that initiate nuclear reactions.

But it is difficult to measure the power in a lightning bolt exactly. Estimates based on how much the bolt heats the air it traverses, and on its electrical properties, give a range of 1 to 10 gigawatts (1 gigawatt = 1 billion watts) over the second or so that a typical bolt lasts; the film *Back to the Future* (1985) got this right when it showed a 1.21 gigawatt lightning bolt turning a DeLorean sports car into a time-travel machine. In more familiar units, a bolt at the high end of the range carries a staggering 13 million horsepower over its brief lifetime.

It is also hard to measure how much energy is transferred to what the lightning hits, but answers are coming from a new approach that studies "lightning fossils" or fulgurites (from the Latin for lightning). These result when lightning heats sand, soil, or rock enough to melt the material and turn it into glass. In 2016, the geoscientists Matthew Pasek and Marc Hurst analyzed 266 hollow cylindrical fulgurites retrieved from a sand mine in Florida, with lengths from centimeters to more than a meter (other fulgurites up to 5 meters long have been found). Using the known heat needed to vitrify silicon dioxide, the main component of sand, the researchers found that a bolt delivered only a small fraction of its power to change sand into glass.

However, the power transferred depends on the target material. In 2017, Jiangzhi Chen and colleagues at the University of Pennsylvania studied fulgurites in granite, where a lightning strike creates a high-pressure shock wave. Here, researchers found, the thunderbolt at a temperature of 100,000°C heats the rock to over 2,000°C within tens of microseconds, melting it to form a layer of glass. This is comparable with the devastating effects of a meteorite impact. Further studies will gauge these effects in other materials and give a basis to design appropriate protections.

Noting the high power that lightning can deliver, scientists have wondered if it could be helpful as well as destructive by providing free renewable energy. At Lake Maracaibo, a single gigawatt harvested from each of the 233 flashes per square kilometer per year could operate 20 residences. This would require building a network of collecting stations to capture the flashes, storing the power surges as they arrive at irregular intervals, and doling out the power to users as needed—a complex engineering project that could not compete economically with other energy sources. But once we learn lightning's ancient secret—what triggers it—maybe we can learn how to keep it from even beginning before it does harm, and how to concentrate it in places where its power can be collected and used.

Whatever questions remain about ordinary lightning, even less is known about the rare and exotic form called ball lightning, a glowing sphere that appears near a lightning strike or thunderstorm and floats through the air for several seconds. One early eyewitness account from 1638 describes a ball of fire more than 2 meters in diameter that entered a church in Devon, England, killing four people and damaging the building. Thousands of other sightings have been reported, and in modern times ball lightning has been seen to penetrate glass and to appear inside closed metal aircraft. Many hypotheses have been proposed to explain this remarkable phenomenon but, with little quantitative data, no definitive explanation has yet emerged.

In 2012, however, Chinese scientists observing ordinary lightning had the good luck to see a glowing ball develop from a nearby ground strike. They recorded photos and videos, and the first ever spectrum of the radiance from ball lightning. This indicated the presence of silicon, a main component of soil. The result supports a theory put forth in 2000 that ball lightning arises when a lightning strike converts silicon in the ground into silicon compounds in nanoparticle form. Projected into the air, these oxidize at a relatively slow rate to generate a characteristic long-lasting glow. The hypothesis is yet to be confirmed.

Despite the enduring mystery, what we do know has helped us study other celestial bodies with an atmosphere, such as Jupiter. This planet displays extensive lightning activity, as first seen by the *Voyager 1* spacecraft in 1979, and observed today by the *Juno* spacecraft that began orbiting Jupiter in 2016. Juno has detected radio waves from hundreds of lightning flashes believed to have arisen from charge separation between water and ice, just as on Earth. But Earthly lightning is densest near the equator, whereas Jovian lightning is concentrated near that planet's poles. This is an important clue toward understanding the distribution of water on Jupiter and the planet's atmospheric dynamics.

Widespread lightning has also been seen on Saturn, which has its own active atmosphere. Surprisingly, Mars—with its thin carbon dioxide atmosphere—displays lightning as well; but in that arid environment, it is "dry lightning" that does not depend on atmospheric water and ice. Instead, as also happens on Earth, these electrical discharges come from friction among the tiny particles carried by the strong dust storms frequently seen on Mars.

Martian lightning is intriguing because the planet is a potential site for past if not present alien life. Starting in 2008, NASA landers have found perchlorates, compounds containing the negative ion ClO_4, in Martian soil. These have attracted attention because they can fuel certain microorganisms that might have lived on ancient Mars. The perchlorates were more abundant than Martian geology would indicate, suggesting that they were made by lightning. Now an international research team has just found that electric discharges in a simulated Martian environment create perchlorates in large amounts, with implications for the evolution of alien life.

This experiment, carried out for a planet named for an ancient god, reminds us both of a time when the gods seemed to rule and of the 1952 Miller–Urey experiment that first sought a scientific explanation for life's beginnings. Lightning embodies humanity's evolution from belief in a universe controlled by the gods to belief in a natural world that we can grasp, with mysteries yet to be solved.

We and the Earth Breathe Together

Reviews of the films *Coronation Park, Deforest, 2042, Smog, Days of Eva, Expire, Plastic Child, Grow, Grassroots*, and *The Ocean Takes a Deep Breath* (these films can be seen at the Labocine site https://www.labocine.com/)

The 33 films in the Labocine September issue "Breath of Life" show in different ways how our planet and we ourselves breathe... or don't. Plants support all Earthly life by using the Sun's power to take in carbon dioxide and expel oxygen. We and other living creatures do the opposite as we breathe in and out to support our life processes. If the grand global carbon dioxide–oxygen exchange were to fail through extreme pollution or climate change, that would harm or completely end us and every other living thing. And when the personal cycle of respiration fails for any individual, that means illness or death; or alternatively, if a person's breathing can be enhanced and controlled, that can improve fitness and bring the serenity of meditation.

These themes and others appear in the films in "Breath of Life." Many show a pessimistic view of the future of global breathing and our own personal respiration, but optimistic notes appear as well. One is the short effort *Coronation Park* (2015, Su Rynard). It celebrates a city park in Toronto, Canada, through images of its bare trees waiting to bud into spring greenery, with the word "breathe" appearing on screen in different languages. Automobiles appear fleetingly to remind us that they spew carbon dioxide, which the trees absorb to release oxygen.

These images are hopeful, but trees alone do not ensure abundant oxygen. Surprisingly, even infinitely huge forests such as the Amazon jungle barely contribute to atmospheric oxygen. Although the Amazon has recently been cited in the media as producing 20% of the oxygen we breathe, that oxygen actually has mostly come from ocean-based plant life. Besides, most of the oxygen from land-based forests goes into other processes, not into the atmosphere. The concern about the fires now burning in the Amazon is rather over the loss of its ability to absorb atmospheric carbon dioxide, a major factor in global warming, and the loss of habitat and a marvelous ecosystem; but not because this or any other deforestation will deplete oxygen. Carlos Nobre, a leading Brazilian climate scientist, put it best when he said that the Amazon is "not really the lungs of the world, no."

But trees and plants represent nature, and with the other reasons to treasure forests, failing to preserve them would cost us dearly as *Deforest* (2015, Grayson Cooke) shows. The film's subtitle "H_2SO_4" is the chemical formula for sulfuric acid, whose extremely corrosive nature can dissolve metals. Cooke uses it instead to dissolve photographic representations of forests, beautiful monochrome slides of trees, branches and the tree canopy within a temperate rainforest in Australia's Bunya Mountains. We watch the acid eat into the slides, making them crack, fold and distort in patterns reminiscent of deserts and harsh geological features rather than green communities of life. Against a soundtrack of real bird and animal sounds, thunder, and rain recorded in the Bunya rainforest, and melancholy piano chords, the film combines the experience of living in nature with a sad appreciation of what its loss means.

Although not explicitly about the effects of air pollution, *Deforest* refers to it through the use of sulfuric acid. That is a component of acid rain, the destructive precipitation that comes from the pollutants released by the burning of fossil fuels. The scientific consensus is that continued use of fossil fuels is leading to the even worse effects of global warming. If we believe the filmmakers who contribute to "Breath of Life," pollution and global warming will also create a breathing apocalypse. Several of the films present fictional views of near-future dystopias where the atmosphere is so toxic or the oxygen level so low that people cannot properly breathe.

Though these films show different stories told in individual styles, they share some characteristics. One is the idea that land-based plants can restore a toxic atmosphere to breathability. Though oxygenation is more complex than that, as I discussed, this is an effective "back-to-nature" symbol. Another is that the films are mostly shot in monochrome and sometimes with fog effects to represent a dark and poisoned atmosphere. A third cinematic and thematic feature in common is that the people in the films wear oxygen masks or gas masks to breathe. Whether a military-style rubber gas mask with a snout-like filtration canister, or a sleeker plastic version connected to an oxygen tank, the characters look alien and dehumanized. We the film viewers cannot read their facial expressions and neither can they among themselves. This social and emotional separation contributes to the societal breakdown that accompanies the atmospheric disasters in the films.

2042 (2010, Emiliano Castro Vizcarra), from Mexico, ascribes that breakdown to a general prophecy said to be from the Mayan culture that could also describe global warming: as time goes on, "man will conspire to destroy this divine cycle [of life]," by first destroying "the plants and the planet" and then himself, leading to a new consciousness. *2042* opens with destruction in full view as gunshots and explosions fill urban streets. A man runs desperately through the chaos and delivers the child he carries to a building protected by armed guards. It is a hospital for children, who must wear gas masks to play outside. A church bell rings and the children gather to watch the unveiling of a wonder, a single green plant, perhaps the only one left. But the final scene undercuts hope as we see the desperate man check his automatic pistol, ready for more destruction.

Other of the dystopic films are more specific about the human causes of atmospheric failure. *Smog* (2014, Jad Sleiman) begins by explaining, "In the near future, factories filled the planet and intoxicated the air. Humans started living in the deserted waste where smog covers the skies and hides the Sun." Shot in monochrome among deserted industrial landscapes and piles of debris, we see 12-year old Lithops trudge through this ugly and sterile world, protected by layers of clothing and a huge gas mask with staring eyepieces. She survives for a while, but with little joy or pleasure and with her once happy life represented only by an old family photo, and a rag doll she finds. Finally, but only in death it seems, Lithops ascends into clean air and sunlight, shot in color for the last seconds of the film.

In *Days of Eva* (2016, Vincent René-Lortie), a young woman wears a mask and carries a big oxygen tank in a world where air has become unbreathable. We meet Eva as her wrist read-out flashes into the red zone, showing only a day's worth of oxygen remaining. She desperately seeks oxygen or any kind of help but cannot find either. Despairing and angry, with all hope gone, she destroys mementos from her past life; then, as the oxygen warning beeps escalate, removes her mask and lies down to die, without even a vision of a more natural world such as Lithops had.

Expire (2017, Magali Magistry) features teen-aged Juliette in a future where people live indoors, fearful of a universal dense smog.

When her boyfriend Mehdi messages, she takes canisters of oxygen and a mask and sets out through the smog. On the way she is attacked by two men in what could be a rape but becomes a robbery when they take her precious canisters and flee. Not seriously hurt, she reaches Mehdi at a party where young people dance while wearing their masks. The two feel the need to pierce the physical separation the masks enforce and remove these barriers so they can kiss, a moment of pure human connection in an inhuman time.

Plastic Child (2016, Carolin Koss) provides a different kind of redemption in an unnatural world. A nameless young boy dressed in a white bodysuit with only his face showing awakens in some unknown place. He puts a tube into his nostrils that is connected to a single living plant that he carries on his back. This allows him to breathe as he crawls, walks, and swims to a world filled with plastic in heaps and covering tree trunks. This is our contaminated future Earth, which he slowly explores until he encounters real grass and a path leading to a mound of soil. He kneels, takes the plant from his back and gently sets it into the soil, then removes his breathing tube.

This symbolic return to a natural balance in the next generation is at odds with the harsh view in *Grow* (2015, Micah Levin), which outlines the chronological development and corporate exploitation of a polluted Earth: 2033, climate change reaches its tipping point; 2045, the Illuminet Corporation takes over New York City; 2053, with air now unbreathable, Illuminet corners the oxygen market, banning plants as alternate sources and executing people who grow them (in the Labocine film *Sleep Dealer* (2008, Alex Rivera), future corporations seize control of another vital resource, water).

Grow begins as Winston Willis, an illegal oxygen dealer, meets a mysterious hooded would-be buyer of oxygen in a seedy bar, much as a drug deal would unfold. Willis is tracked by an Illuminet drone, and as he negotiates and then grapples with the buyer, violence breaks out with the participation of the Free Breathers, who rebel against corporate control of air. Willis escapes from the bar, but as the film ends, we know only that the battle for control of breathing will go on.

These fictional films properly raise the alarm about the apocalyptic future of our planet's atmosphere. They offer no solutions beyond appreciating the importance of nature, but the documentary *Grassroots* (2018, Frank Oly) does. Set in Australia, its title refers both to individual "grassroots" participation in battling climate change, and to a battle plan that uses the actual roots of growing plants.

The film shows this through images, narration, and interviews, beginning with retired University of Sydney professor Peter McGee. In 2012, he lectured about his research on a particular fungus that thrives in the roots of plants where it does something special: it takes carbon from the air where there is too much, and puts it into the ground where there is too little, as explained by Guy Webb, a soil management expert and the story's leading figure. This is a double win. Putting more carbon into the soil would be good for plants and therefore for farmers, much as the related method of nitrogen fixing has improved global crops over the last century; and locking away atmospheric carbon in the earth would reduce atmospheric carbon dioxide and therefore inhibit global warming.

These prospects inspired people like Webb, who saw the potential for a global army of two billion farmers to increase crop yields while sequestering gigatons of atmospheric carbon. With colleagues, he has raised money for initial research, but knows that more minds need to be changed and more people need to be motivated to have a wide impact. At film's end, McGee, who started the whole thing, says that though belief in climate change is growing, "I deal with politicians who don't care what's happening to the globe;" an Australian farmer points out that the mysterious "they" who will solve global warming is really us, all of us; and Webb says simply that we need to accept that there is a real issue and "ruddy well solve it." This is a key message from *Grassroots*: besides scientific tools to reverse climate change, we need widespread political and social will.

Besides, there is still much we don't know about breathing on the global and the individual scale, as other films in "Breath of Life" show. For instance, the animated documentary *The Ocean Takes a Deep Breath* (2017, Saskia Madlener) explains that the Earth's seas participate in global breathing in ways we don't yet understand. The film focuses on the Labrador Sea, between Labrador in Eastern Canada and Greenland. In this particular region, oxygen and carbon dioxide are carried to great depths and sequestered. Researchers are now probing how this affects sea life, the climate, and global warming. Other films display how the lungs work, the role of breathing in personal meditation, and more. Finally, "Breath of Life" is a quick course in the importance of respiration in the world and in ourselves.

The Better to See You With

The first time I had my heart checked by ultrasound at the Emory Clinic, I didn't know what to expect. The technologist pressed a small device against my chest and moved it around, using a gel-type substance as a conductor, until an image appeared on a computer screen off to the side. Craning my neck, I could just make out a moving shape. It took me a second to realize what I was seeing. It was my own heart, beating even as I watched it—the closest I've ever come to an out-of-body moment.

As a physicist, I understand the mechanics of medical imaging. Put simply, images from inside your own body are gathered using sound waves, X-rays, gamma rays, or radio waves and a magnetic field. These physical probes enable ultrasound, X-ray computed tomography (CT), positron emission tomography (PET), and magnetic resonance imaging (MRI)—medical tools developed by physicists, engineers, and physicians that are now essential to patient care.

Jon Lewin, executive vice president for health affairs at Emory and CEO of Emory Healthcare, is a professor of radiology and imaging sciences and biomedical engineering, and an innovator in the field. What drew Lewin to medical imaging was its trifecta of tech, biomedical science, and patient impact. "When I got involved in the mid-1980s," he says, "imaging was going from being very static—looking at X-rays, shadows on a screen—to interrogating and visualizing the human body as never before, even during surgery."

And the level of detail just keeps accelerating, he says. "The kind of information it is possible to learn through current imaging is transforming how medical care is provided across almost every specialty." Take, for example, functional brain imaging—"how thoughts are translated into electrical impulses and blood flow," says Lewin. Or the advances made possible by interventional radiology. "Minimally invasive, image-guided therapies have changed what used to be major surgery and a week in the hospital into outpatient surgery through a needle hole," Lewin says. "I had to tell one patient to at least take the weekend to recuperate."

The dark ages

Long before these techniques existed, physicians had to try to heal patients without knowing much about what was inside their bodies. Anatomists and physicians eventually learned more through invasive surgery, by dissecting dead animals and people, and by using tools like the microscope. But it took an accidental observation in 1895 to allow doctors to finally peer inside the living human body without cutting into it.

German physicist Wilhelm Roentgen was working with a novel lab device, a glass tube pumped free of most of its air, with a metal electrode at each end. When high voltage was applied to the electrodes, a glow appeared between them. That was not new, but something else was. Though the tube had an opaque covering, Roentgen noticed that whenever it was turned on, a fluorescent screen three meters away would light up. Roentgen had discovered an unknown kind of penetrating radiation he dubbed "X-rays." They generated universal wonderment.

Only later did physicists learn that X-rays are energetic electromagnetic waves emitted by electrons streaming within the tube. The medical value of these new rays became immediately clear when Roentgen displayed an X-ray photo of the bones in his wife's hand. In less than a year, American physician William Morton published *The X-Ray: or, Photography of the Invisible and Its Value in Surgery* as a guide to medical use of the technique.

From 2D to 3D

The early medical use of X-rays did much good but also harm.

With their high energies, X-rays penetrate the soft flesh of the body until they are absorbed by the denser bones, which appear as shadow images in an X-ray photo. Those high energies also can damage DNA and harm living cells. Before this was understood, X-rays were indiscriminately applied. As late as the 1970s, they were used in shoe stores to show a customer's foot within a shoe. X-rays were also causing cancer in some researchers who worked with them. Now we carefully limit X-ray dosages for practitioners and patients. Still, the potential harm from X-rays highlights the need for alternative imaging methods with much lower potential for tissue damage.

Computed tomography (CT) creates X-ray images of cross sections of the body that a computer then assembles into a detailed image. When this is done from different angles, changes in density can be converted into pictures of soft tissue, along with bones and blood vessels. CT can diagnose conditions from cancer and cardiovascular disease to spinal problems and trauma.

At Emory, biomedical engineer Amir Pourmorteza, assistant professor of radiology and imaging sciences, and colleagues have replaced the usual devices that detect X-rays in CT with more sensitive and selective "photon counting" units. The group's recent pre-clinical trials on human subjects demonstrate that these detectors enhance images of the carotid arteries that carry blood to the head and brain, while reducing X-ray dosages by more than 30%.

"CT scanners have become lower in radiation, faster, and more precise in two ways: less noise (grainy appearance) and higher image resolution," Pourmorteza says. "We're hoping to double resolution." Think of it like TVs or digital cameras, he says—the same improvement in resolution has taken place in CT scans as well. "You can detect tumors sooner. And you can actually see different colors now, which helps us differentiate diseases," he says. "It's like trying to separate an apple from the background leaves—if they both look grey, it can be difficult. But if you can separate the red and green, you can see the apples and the leaves, or the differences between two dense tumors."

Xiangyang Tang, associate professor of radiology and imaging sciences and director of the Laboratory of Translational Research in CT, works with other X-ray detectors and computer algorithms to turn CT data into detailed images of organs like the heart, potentially at lower X-ray exposures. "Each modality has its own advantages and disadvantages. Our responsibility, as radiologists, is to make sure we do the right thing for the patient," Tang says. "One of the most exciting progressions in CT is in cardiac imaging. Rapid heartbeats used to cause blurring, but increased scanner speeds mean physicians don't have to administer beta-blockers to reduce a patient's heart rate. The new technology can cover the whole heart in one rotation."

Mutual annihilation

Like CT, positron emission tomography (PET) also assembles 2D images into 3D, in this case to locate normal and abnormal metabolic processes in the body.

Positrons are elementary particles first suggested in 1928 when French physicist Paul Dirac predicted that an electron has a kind of fraternal twin, exactly the same but with the opposite (positive) electrical charge. This was confirmed when positrons were found in cosmic rays. Now, of course, we know that every kind of elementary particle has an antiparticle and that antimatter exists along with ordinary matter. Antimatter is a staple of science fiction because of a dramatic fact: when antimatter meets matter, they annihilate each other in a burst of energy. That burst makes PET the only practical application of antimatter so far.

In PET, the patient is injected with a biologically active and weakly radioactive compound, frequently a special formulation of the sugar glucose. As this is taken up by bodily tissues, it emits positrons, each traveling a millimeter or less before meeting an electron resident in the body. The particles mutually annihilate and generate two gamma rays, powerful electromagnetic waves that, like X-rays, must be carefully monitored. The gamma rays travel in opposite directions until they are detected outside the body, at which point a computer tracks their paths back to their internal point of origin. Adding up many such calculations produces images of regions with strong metabolic activity, as shown by their glucose uptake.

A high metabolic rate typically shows the presence of a tumor, so a main use of PET is to identify and monitor cancers. Brain activity also involves glucose metabolism so PET can diagnose brain tumors, guide brain surgery, and confirm diagnosis of brain deterioration, as in Alzheimer's disease. Carolyn Meltzer, chair of radiology at Emory, specializes in PET, and has shown that the best results come when a patient is evaluated by a combined PET/CT scanner—the resulting near-perfect alignment of the two sets of images outperforms separate scans. For instance, when squamous cell carcinoma, the second-most prevalent type of skin cancer, is suspected in the head and neck, a PET/CT scan facilitates the ability to differentiate between cancer and other abnormalities.

Radiology has medical applications far beyond what people normally think of, says Meltzer. "We are interested in expanding work with imaging agents that have both diagnostic and therapeutic functions, as well as further innovation in image-guided interventional treatments for such wide-ranging conditions as obesity, pain, and tumor ablation."

Alternative to opioids

Interventional radiology (IR) is, indeed, a field to watch. Image-guided ablation is an effective treatment for early kidney cancer. Interventional radiologist Sherif Nour also has performed many successful ablations of liver and prostate cancers at Emory. Last year, he traveled to Egypt to teach the technique. And David Prologo, associate professor of radiology and imaging sciences and an interventional radiologist at Emory Johns Creek Hospital, is helping to develop an IR program in Tanzania.

Prologo is also investigating the use of IR in reducing obesity by freezing a portion of the vagus nerve that sends hunger signals to the brain. But much of his and other interventional radiologists' work is pain related. "We are using our skill set to solve new, long-standing problems, like chronic pain, phantom limb pain, cancer pain. We are able to make dramatic changes in patients' quality of life," he says. One of Prologo's patients had metastatic lung cancer in her spine and intractable nerve-related pain for more than a year. She underwent CT guided cryoablation of the cancer and the nerve, to block the pain signals. "I ran into her at Publix and she was crying because she was so relieved," he says.

IR is being used more and more as a drug-free alternative to opioids for pain relief. "We do our procedure while viewing the CT on a screen, in real time," Prologo says. "It allows us to guide a needle in safely. We inject the nerve with the anesthetic bupivacaine, a temporary block, and if it works on the patient, we feel justified in doing something more permanent, such as ablation." Interventional radiologists work with doctors from all subspecialties—surgeons, ob/gyns, generalists, orthopedists, and bariatric, pain, and palliative care physicians. "We can do more for the patient," says Prologo, "when we all work together."

Like tiny magnets

The radiation exposure that CT and PET scans require can be an issue for pregnant women or for patients who require frequent scans for diagnosis or follow-up. That's where ultrasound and magnetic resonance imaging (MRI) come in. MRI grew out of nuclear magnetic resonance, invented in 1939 by the American physicist Isidor Rabi to study atomic nuclei.

Like tiny magnets, nuclei placed in a strong magnetic field align themselves in specific configurations with different energies. When excited by radio waves, the nuclei emit their own electromagnetic waves at frequencies corresponding to their energies, which act as markers for the particular type of nucleus. This evolved into the biomedical MRI, which exposes the patient to electromagnetic radiation in the form of radio waves that are too weak to harm cells. The scan also requires a magnetic field thousands of times stronger than the Earth's. This normally does no harm, although MRIs cannot be used on patients with metallic implants, such as pacemakers.

MRI is a versatile tool for diagnosis and treatment of the neural, cardiovascular, musculoskeletal, and GI systems. For example, two different MRI studies of heart patients, led by John Oshinski, interim director of the Emory Center for Systems Imaging, showed how to best place the electrical lead from a pacemaker into the wall of a heart to control its beating, and evaluated the effectiveness of a technique to treat rapid, irregular heartbeats.

MRI can also assist in taking research from bench to bedside: for example, the use of stem cells and nanomedicine. In 2016, Oshinski, an associate professor in radiology and imaging sciences, and co-researchers used MRI to track stem cells injected into the spinal cords of pigs. The cells were tagged with nanometer-sized iron oxide particles that interacted with the magnetic field during the MRI. They appeared in the resulting images where they confirmed that the cells were properly placed and continuing to function. Imaging will be critical to develop therapeutic methods like this for the central nervous system. Magnetic nanoparticles also played roles in recent MRI investigations by Hui Mao, professor of radiology and imaging sciences, showing how nanoparticles carry drugs into tumors and other targeted locations.

My heart will go on

Medical ultrasound differs from the other techniques in using sound waves, not electromagnetic energy or nuclear processes. It began with the development of methods to detect underwater objects after the *Titanic* hit an iceberg and sank in 1912, and to counter German U-boats during World War I. In response, the French physicist Paul Langevin invented devices to put sound waves into water and register the echoes as the waves are reflected from objects, similar to echolocation as used by dolphins and bats. The result was sonar.

Sonar, in turn, inspired the Scottish gynecologist and obstetrician Ian Donald, who learned about it during his military service. In the 1950s, improvising with industrial ultrasound units that detect flaws in metal, he imaged ovarian cysts in women and published a paper showing the first pictures of a fetus's head in a pregnant woman. Commercial medical ultrasound units soon followed.

The sound waves used in medical ultrasound have frequencies far beyond the upper limit of human hearing, 20 kilohertz. They are typically in the range 1 to 18 megahertz, with the higher frequencies sensing finer details and the lower values providing deeper penetration into the body. A vibrating transducer—the device the technologist at Emory used to examine my heart—is placed against the body or sometimes in a body cavity to send out ultrasound waves. Like light waves, these are reflected, scattered, refracted, or absorbed as they encounter organs and tissues. The returning echoes are detected, and a computer analyzes the results to form a visual image of what the sound is passing through.

With its lack of damaging radiation, obstetric ultrasound is widely and safely used to examine the fetus in pregnant women, allowing prospective parents to see inside the womb. The portability of ultrasound equipment and its real-time results mean it can quickly image tendons, muscles, joints, blood vessels, and organs. For instance, it can be used to examine the size, shape, and pumping capacity of the heart right in the doctor's office.

The sound and the fury

Successful ultrasound imaging depends on putting sound energy into the body with efficient transducers and obtaining high-contrast images. These can be enhanced by using contrast agents—gas-filled microbubbles that strongly reflect sound waves and improve the contrast when blood vessels are being examined. Researchers at the Coulter Department of Biomedical Engineering (BME) at Emory and Georgia Tech address both of these issues. Brooks Lindsey, assistant professor of BME, is using Georgia Tech's clean room facilities and other resources to design better transducers and contrast agents in cooperation with Emory physicians.

And Stanislav Emelianov, professor of electrical and computer engineering and BME and director of the ultrasound imaging and therapeutics research lab, has developed a new laser-based contrast agent. In 2017, his team showed that nanodroplets filled with a specific liquid can be vaporized by laser pulses to form highly reflective, gas-filled microbubbles. These quickly re-condense into liquid bubbles that can be reactivated over multiple cycles. This method produces superior high-contrast images of lymph nodes, which can show the spread of cancer.

"The radiology department provides access to imaging for researchers throughout Emory, not only clinical, but also brain health, public health, and animal research," says Elizabeth Krupinski, professor and vice chair for research in radiology and imaging sciences. "A lot of key studies in developing biomarkers, contrast agents, and tracers have to be done in animals first."

The future of imaging science will undoubtedly involve artificial intelligence and big data. But imaging can never completely take the place of good clinical skills, says Laurence Sperling, who directs Emory's preventive cardiology program. Sperling reminds students and young physicians that effective clinical care requires "listening to, talking to, and touching the patient. We are taking care of people, not pictures of people."

The Power of Crossed Brain Wires

When I was about 6, my mind did something wondrous, although it felt perfectly natural at the time. When I encountered the name of any day of the week, I automatically associated it with a color or a pattern, always the same one, as if the word embodied the shade. Sunday was dark maroon, Wednesday a sunshiny golden yellow, and Friday a deep green. Saturday was interestingly different. That day evoked in my mind's eye a pattern of shifting and overlapping circular forms in shades of silver and gray, like bubbles in a glass of sparkling water.

Without knowing it, I was living the unusual mental state called synesthesia, aptly described by synesthesia researcher Julia Simner as a "condition in which ordinary activities trigger extraordinary experiences." More exactly, it is a neurological event where excitation of one of the five senses arouses a simultaneous reaction in another sense or senses (the Greek roots for "synesthesia," also spelled "synaesthesia," translate as "joined perception"). Some 4% of the population experiences this kind of cross-sensory linking, and studies have shown it's more prevalent in creative people. Artists who've reported extraordinary experiences of synesthesia range from 19th-century composer Nikolai Rimsky-Korsakov to contemporary artist David Hockney to pop music star Lady Gaga.

For me, the words "Sunday," "Monday," and so on, generated internal visions of color and pattern. Most synesthetic reactions also involve color in response to lexical stimuli—words written or spoken ("word-color" synesthesia), and letters, numbers, and symbols ("grapheme-color" synesthesia)—or to music and sound ("colored-hearing" synesthesia). Researchers have also observed dozens of other types of stimulus-reaction combinations: taste evoking a visual image, such as the flavor of chicken producing a 3D shape; physical touch inducing the sensation of smell; and somehow the most extraordinary pairing, words generating the sensation of taste, such as "jail" creating the flavor of bacon.

At one time, strange pairings like these were little understood and even feared as signs of pathology. Now we know more, but many questions remain and synesthesia still carries an exotic aura. At the same time, it is a new tool to explore the human brain and mind, creativity, and consciousness.

Reports of sensory cross-connections go far back. In 1690, the English philosopher John Locke wrote of a blind man who associated the sound of a trumpet with the color scarlet, although it is unclear if this was synesthesia or a metaphor, a recurring issue in synesthesia research. In 1812, however, a German physician wrote a definitive description of seeing colored letters. Other physicians reported similar experiences in patients and the reports drew attention from scientists, clinicians, and artists. French poets Arthur Rimbaud and Charles Baudelaire extolled the romantic idea that the senses should intermingle.

But 19th-century scientific and clinical understanding was limited, and synesthetes were often reluctant to come out of the closet for fear of appearing "odd" or worse. Some synesthetes were diagnosed with conditions such as schizophrenia when their cross-sensory effects were taken as delusions or hallucinations. Or clinicians denied that synesthesia existed, interpreting patient's statements like this "music looks red" as over-enthusiastic metaphor.

Fortunately, the scientific study of synesthesia grew from the late 19th century into the 20th, mainly using interviews and group surveys. In 1895, psychologist Mary Whiton Calkins at Wellesley College analyzed her students' self-reported responses about synesthetic manifestations and speculated about their causes. But like every other scientist at the time, she had no way to probe the phenomenon at a deep neural level. Nevertheless, research continued and led to several international conferences on the subject in the 1930s.

But by then, research had diminished. One problem was the difficulty of tracking the many forms of synesthesia. Another is the same reason that human consciousness is hard to study—the synesthetic experience is internal and subjective. It's problematic to analyze a neural event when the only evidence is the subject's personal account. The entire field of psychology recognized this in that same era when it turned away from studying internal experiences.

In the 1980s, however, new approaches made synesthesia amenable to more rigorous and objective study and research has blossomed, with about 1000 new publications since 2000. One big step has been the acceptance of a consensus definition of synesthesia. Its key features are that a subject involuntarily experiences vivid responses to stimuli that combine two or more different sensory modalities, and that the responses are constant over time; for instance, a given word always induces the same color in a particular subject (the colors themselves are unique to each person). The psychologist Simon Baron-Cohen and colleagues introduced this last benchmark in 1987 as an objectively measurable standard of genuineness. True synesthetes give the same responses to the same stimuli when tested and retested over long time intervals. Childhood synesthesia generally continues into adulthood, though not always. I lost my automatic color associations by the time I was 12. Today I can only remember the colors.

Besides these tests, new neuroimaging techniques have established synesthesia as a real neurological process. One widely used method is functional magnetic resonance imaging (fMRI) of the brain. Unlike regular MRI, which shows the anatomy of the brain (or other internal organs), fMRI identifies which parts of the brain are active, nearly in real time. Since 2002, some fMRI investigations of grapheme-color synesthesia—the most widely studied kind—have shown that graphemes stimulate the V4 region of the brain. This area deals specifically with color within the visual cortex, the part of the brain that processes what the eyes see (the auditory cortex and other specialized areas handle the remaining senses). This is consistent with a model where the regions of the brain that analyze graphemes and that deal with color are somehow hyper-connected to create a synesthetic event. But not all fMRI data show the same result, nor is it clear if it is the visual perception of a grapheme or its conceptual meaning that is the trigger [1].

Remarkably, even with advanced methods, other basic questions have lingered since the 19th century. How prevalent is synesthesia in the general populace? Different studies had given values from over 20% to less than 1%, a disparity partly due to the use of self-reported data. To remedy this, in 2006 Julia Simner, then at the University of Edinburgh, and her colleagues carried out a controlled approach. They interviewed nearly 1,700 subjects and tested them for consistency over time. The certified synesthetes constituted 4.4% of the group, 1 for every 23 people—rare, but not vanishingly so. The data also showed that the most common subvariant is the one I experienced: colored days of the week.

Surveys and tests illuminate possible connections between synesthesia and artistic or creative ability. Catherine Mulvenna at University College London, who has written about this elusive connection, asks, is synesthesia "a driving force or a mere idiosyncratic quirk" in artists? As Mulvenna points out, one connection of synesthesia with creativity is personal testimony from creative people. This often comes across as compelling evidence of true synesthesia. Vladimir Nabokov wrote in his autobiography *Speak, Memory* of his "colored hearing," in which the letters x and k are respectively a steely blue and a huckleberry blue. Richard Feynman has related, "When I see equations, I see the letters in colors ... light-tan *j* s, slightly violet-bluish *n* s, and dark brown *x* s flying around." Hockney sees colors when he hears music, which he uses when he does stage design for ballets and operas.

Many people with musical talent give detailed accounts of how music makes colors for them. Besides Rimsky-Korsakov, Jean Sibelius feared mockery for revealing his synesthesia, as did violinist Itzhak Perlman. In the worlds of jazz and popular music, Duke Ellington sensed both colors and textures from music, seeing dark blue burlap or light blue satin for specific notes played by certain musicians; the late jazz pianist Marian McPartland saw the key of D as daffodil yellow and B major as maroon; and Lady Gaga said in an interview, "I do hear music all at once and in lots of colors. It's like a painting."

Research supports these accounts. Self-reported synesthetes appear at a relatively high rate among artistic types, and one study using objective testing found 7% synesthetes among 99 art students compared to 2% in a control group. There is also evidence of associations between synesthesia and elements of creativity, but one stumbling block is the lack of a satisfactory definition of creativity. As Mulvenna writes, "Creativity is a complex construct to define, almost to the point of infamy within fields of systematic investigation."

Nevertheless, researchers have uncovered suggestive correlations between synesthesia and creativity, at least as defined by psychological testing. In 2016, Charlotte Chun and Jean-Michel Hupe, at the University of Toulouse, compared a selected group of 29 artistic synesthetes to 36 controls. Testing for aspects of creativity and related abilities such as mental imagery showed higher scores for the synesthetes, but with smaller differences than in earlier studies. These outcomes and other research results point to qualities that synesthetes consistently display, which may explain their creative tendencies: cognitive traits such as a disposition to think in images and sensitivity to color; and the psychological personality trait called "openness to experience," exemplified by intellectual curiosity and an active imagination.

Examining the synesthesia-creativity link is an important long-term effort, but there is a deeper question for synesthesia itself: What are its roots? In 2018, neuroscientists Simon Fisher and Amanda Tilot at the Max Planck Institute for Psycholinguistics, Nijmegen, the Netherlands, and colleagues published genetic data for synesthetes from three unrelated families where the condition is prevalent [2]. Earlier work had not found individual genes that are responsible for synesthesia, but the new researchers cast a wide net. They obtained DNA from four to five synesthetes and at least one non-synesthete from each family, spread over generations. All the synesthetes displayed the colored-hearing variant called "sound-color."

Using gene sequencing, the researchers identified 37 genes that indicate a tendency toward synesthesia. No single gene was seen in all three families, confirming that the inheritance of synesthesia is not tied to just one gene or group of them. But 6 of the 37 genes were linked to the development of axons, the long thin part of a neuron that communicates with other brain neurons through electrical impulses. Equally intriguing, the six genes are also linked to the early childhood development of the visual and auditory cortices, and the parietal cortex, the "association area" of the brain that integrates information across sensory modes for effective overall functioning.

Since then, Fisher and Tilot have extended their new genetic understanding to study links between synesthesia and other neural conditions [3]. The genetic results and these overlaps are important because, as Fisher told me in an email, they "give us insights into shared neurobiological mechanisms ... increasing our understanding of how synesthesia develops." The genetic links to the development of the sensory cortices of the brain, along with fMRI brain scanning results, seem to support a theory based on hyper-connectivity among brain regions as a basis for synesthesia. More complete genetic data may also ultimately show why the synesthesia genes have survived the evolutionary process, perhaps because they offer adaptive advantages such as seeding creativity.

Synesthesia researchers have now entered the final frontier of brain studies: consciousness. They believe a new understanding of synesthesia can help solve the hard problem of how subjective experience can arise from the physical brain. Myrto Mylopoulos, a philosopher of mind at Carleton University in Canada, and Tony Ro, a neuroscientist at the City University of New York, have written that synesthesia research can broadly test theories of consciousness because the varied forms of synesthesia involve all the senses and cognitive elements [4]. The authors, with other researchers, believe that the empirical approach to finding or confirming a theory of consciousness is to examine its neural correlates, the objectively testable brain functions that must accompany subjective experiences.

Pursuing this idea, Mylopoulos and Ro considered how synesthesia can act as a test bed to choose among theories of consciousness, and examined prevalent candidates, none of which is as yet supported by much empirical evidence. "Higher order" theory assumes that conscious states are those that a person is aware of being in, which comes from another mental state operating at a higher level; but in "first-order" theory there is no need for a higher state because even a perceptual state such as viewing a flower is considered to be a conscious state. Significantly, these differences produce characteristic neural correlates operating in different parts of the brain for each theory. The authors conclude that the scope of synesthetic events spread over the brain and its functions can yield "initial clues as to the neural correlates of conscious perceptual experiences more generally," and help guide researchers toward a valid theory of consciousness.

As for me, I miss seeing the vivid colors of the days of the week that once enriched my internal vision. The colors now exist only in memory. But I would like to believe that synesthesia's link to creativity has enlarged my own mind, especially in my career-long devotion as a scientist and writer to interdisciplinary work. Even under strict scientific study, and centuries after its first observations, synesthesia retains its power to make ordinary life both more marvelous and more complex.

References

1. For a recent overview of synesthesia research, see J. Ward and J. Simner, Synesthesia: The current state of the field, in Sathian, K. and Ramachandran, V. S. (eds.) *Multisensory Perception: From Laboratory to Clinic.* Academic Press, London, U. K. (2019).

2. A. K. Tilot et al. Rare variants in axonogenesis genes connect three families with sound-color synesthesia. *Proceedings of the National Academy of Sciences*, 115 (2018), pp. 3168–3173.

3. A. K. Tilot et al. Investigating genetic links between grapheme–colour synaesthesia and neuropsychiatric traits. *Philosophical Transactions of the Royal Society* B 374, 20190026 (2019). See also https://www.mpi.nl/page/join-our-synaesthesia-genetics-research.

4. M. Mylopoulos and T. Ro, Synesthesia and consciousness: Exploring the connections, in O. Deroy (ed.), *Sensory Blending: On Synaesthesia & Related Phenomena.* Oxford University Press, Oxford, U. K. (2017).

Can Zapping Your Brain Really Make You Smarter?

According to the advertising hype, you too can enjoy incredible neural and psychological benefits in the comfort of your own home by using a simple electrical device that offers transcranial direct current stimulation (tDCS). For instance, three different models of tDCS devices sold online claim to improve mood, increase creativity, enhance memory, accelerate learning, and combat pain and depression. For the low, low price of between $99 and $189.95, you get a compact handheld device with easy-to-use controls and two electrical leads that end in small sponges. These sponges are dipped into saline solution to make them current-carrying electrodes, then placed against your head. The websites typically show the sponges located on either side of the forehead, but point out that they need to be placed elsewhere on the skull to activate different parts of the brain, depending on the desired outcome.

Wherever you place the sponges, when you switch on the unit, you're pumping electric current into your brain, although not very much. The supposedly safe maximum of 0.002 ampere, as set by a tDCS device, is only a tiny fraction of typical household current. But then, the 86 billion neurons in your brain communicating with each other via electrochemical pulses don't use much current either. Two milliamps (0.002 amps) is enough to change some neural interactions and affect brain function, a result that was firmly established in 2000 [1]. That makes tDCS an important tool for neuroscience. But, despite the glowing user testimonials, the benefits from using tDCS at home are far from definite.

All of the tDCS websites state that the units are not sold as medical devices or to treat medical conditions, and only to enhance "wellness." This exempts tDCS devices from U. S. Food and Drug Administration (FDA) oversight for medical effectiveness (much in the way dietary supplements that claim only to improve wellness are not necessarily subject to FDA oversight).

Without regulation, there is no guarantee that the devices are correctly designed and built. One model tested in a university lab actually impaired memory [2]. Incorrect use and placement of the electrodes can also produce unreliable results. All this could diminish any good effects from tDCS. On the other hand, a user's perception of positive results may be influenced by the placebo effect, where belief in the efficacy of the treatment becomes self-fulfilling. Moreover, tDCS can produce bad or uncomfortable outcomes. Using too much current or simply using the device excessively can be damaging. Besides the burns that some users report, some people experience neural phenomena, such as apparent flashes of light.

Neuroscientists, psychiatrists, and medical ethicists have considered [3] the benefits and risks of electrical brain stimulation, with particular concerns [4] about direct-to-consumer tDCS. Still, people are eager to try it, as therapy or as brain booster, as shown in a tDCS Reddit forum where more than 10,000 members trade tips [5].

Anyone with basic electronics training can build the device at home for little cost. Combine this easy availability with people's faith in technology, and you can see why do-it-yourself tDCS thrives. An aging population worried about declining cognition also contributes to an atmosphere where tDCS, "brain games" to train the mind, and so on, evoke hope.

This contemporary use of electrical neurotechnology follows in a long tradition of sparking the brain with electricity. In the first century CE, Scribonius Largus, physician to the Roman emperor Claudius, described the therapeutic value of a stingray-like fish called a torpedo that delivers electric shocks. Even an unbearable headache, he wrote, "is taken away and remedied forever by a live black torpedo placed on the spot which is in pain" [6].

Later, the eighteenth century Italian researcher Luigi Galvani seriously investigated bioelectricity by observing frog legs twitching under electrical voltage. This soon elicited both pseudo-medicinal and medicinal uses. Amazed audiences watched Galvani's nephew Giovanni Aldini seemingly animate dead people with electrical jolts, though he was only producing temporary muscle spasms. But in 1801, demonstrating perhaps the clearest forerunner to tDCS, Aldini reportedly cured a sufferer of "melancholy madness"—what we would today call major depression—with current from a galvanic battery [7].

Nineteenth century medicine went on to better understand the brain by recording its electrical activity, leading to today's electroencephalograms (EEGs). Then 20th century neuroscience and psychiatry adopted actual therapeutic electrical interventions in the brain. One early approach treated mental disorders with electrically induced convulsions. Known first as electroshock therapy and now as electroconvulsive therapy (ECT), the method has evolved to be less of an ordeal for patients than in Ken Kesey's *One Flew Over the Cuckoo's Nest.* ECT is at least temporarily effective for many people with major depression, but remains controversial and still carries a risk of memory loss. [3]

Deep brain stimulation (DBS), where a "brain pacemaker" implanted in the chest produces electric pulses (like a heart pacemaker), came later. Sent to specific regions within the brain through surgically implanted electrical leads, these pulses mitigate different aspects of Parkinson's disease, a nervous system disorder that produces difficulty in movement, followed by dementia and depression. The FDA approved DBS to treat motor symptoms in Parkinson's in 1997, and hundreds of thousands of brain pacemakers have been implanted since. After testing, in 2018, DBS was also approved by the FDA to treat epilepsy when medication fails, and was first used on a patient in 2019.

Compared to brain surgery or memory loss, tDCS is non-invasive and heals without cognitive damage. It is a research tool as well, allowing scientists to study how specific neural processes cause human behavior. Until now, this fundamental linkage has been investigated only in animal studies, using invasive methods. With proper controls, however, tDCS can be used in research on people, providing new knowledge and a basis for treatments of various neurological and psychiatric conditions. These possibilities were grasped soon after early work on tDCS, in 2003, when it was already being listed as a possible neurotherapy by *Scientific American* [8]. Since then, tDCS research has exploded, with around 5000 scientific and clinical papers published in the last 10 years.

The results so far show that tDCS has therapeutic value in certain areas, when properly used. In 2017, Jean-Pascal Lefaucheur, of Paris Est Créteil University in France, along with an international panel of scientists and physicians, reviewed hundreds of clinical trials that tested tDCS therapy for 13 pathological conditions [9]. The panel concluded that tDCS is "probably effective" for major depression and for addiction to alcohol, drugs and smoking, and "possibly effective" for some types of pain. Otherwise, the trial results were too inconclusive to justify recommendations about using tDCS for other conditions, including Parkinson's disease, Alzheimer's disease, epilepsy, and schizophrenia. But tDCS has given enough positive results that it is now approved for some therapeutic applications in the European Union, Canada, and elsewhere [10].

As is so often the case in science, more work is needed to fully understand any cognitive benefits from tDCS. In 2015, a review of dozens of trials from different research groups, which exposed healthy adults to a tDCS session, showed no conclusive evidence of any effect in four cognitive categories [11]. These are executive function, language, memory, and miscellaneous. Other research, however, suggests that tDCS can improve specific capabilities when correctly applied.

The U. S. Air Force has supported one such study, which is aimed at enhancing human performance in situations like remotely piloting a drone aircraft [12]. This requires dealing with a flood of information over a long duty shift, when alertness can flag. Crucially, the researchers, from Wright-Patterson Air Force Base and Wright State University, carefully placed research-grade tDCS electrodes at specific locations known to affect multi-tasking capability. To guard against placebo effects, half of the 20 Air Force volunteers unknowingly received "sham" stimulation that only mimicked tDCS. In comparison, the real tDCS group showed substantial gains in performing tasks requiring sustained attention.

When PBS science reporter Miles O'Brien underwent the procedure for a television segment, he said that his brain "seemed to turn on like a light bulb" and that for hours after, "It was like a jolt of caffeine without the tense feeling" [13].

To facilitate the further development of tDCS, Lefaucheur and co-authors (and two other authoritative reviews) point to needed changes in tDCS research. They call for a better understanding of the neural mechanisms behind tDCS and of individual differences in the responses to it; for clinical trials to be carried out at multiple research centers, involving more subjects per trial than previous studies; and for consistency in the research protocols used, including controlling for placebo effects, to make comparisons among trials more meaningful.

These measures will focus tDCS research and its transition to clinical use, but do-it-yourself tDCS remains questionable. Lefaucheur and co-authors see value in using tDCS therapy at home rather than at a medical center. It's easier for patients, but Lefaucheur's team only recommends it if the home use is overseen by some form of remote monitoring. However, the authors add, for unregulated and unmonitored home use, the broad availability of tDCS devices makes it unlikely that their safety can be guaranteed.

This is a mounting problem as consumer use of tDCS and of other neurotech applications, such as personal EEG devices and brain fitness software, rapidly grows. Medical ethicist Anna Wexler and psychiatrist Peter Reiner recently noted that this technology is attracting venture capitalists and big corporations [14]. This market segment will likely exceed $3 billion in sales by 2020, making it essential to reliably evaluate consumer neurotech. Wexler and Reiner urge the creation of an independent interdisciplinary agency that would critique the efficacy and risk of neurotech applications without rating individual products. The results would go to consumer groups, health organizations, and regulatory agencies. This oversight group would also consider the ethical and social implications of devices that alter brains [15].

It is hugely tempting to believe in the wondrous cures and enhancements promised for at-home tDCS, but we should remember an earlier rush to new medical technology without proper safeguards that proved harmful. That is the cautionary tale [16, 17] of the rise and fall of Silicon Valley's Theranos company, which was evaluated at $9 billion for a novel blood testing method that proved to be invalid. In the nineteenth century, entrepreneurs falsely sold "snake oil" elixirs as medical cure-alls. Today we need to separate neuro-snake oil from real neural healing and enhancement. We have the tools to do so. We should use them.

References

These references can be found at their individual sources or at the JSTOR digital archive at https://www.jstor.org/.

1. M. A. Nitsche and W. Paulus, Excitability changes induced in the human motor cortex by weak transcranial direct current stimulation, *Journal of Physiology*, Vol. 527, Pt. 3 (Sep 15 2000), pp. 633–639.

2. L. Steenbergen et al., "Unfocus" on focus: Commercial tDCS headset impairs working memory, *Experimental Brain Research*, Vol. 234 (2016), pp. 637–643.

3. K. Hoy, Jumpstart our brains: Hype or hope? *Australian Quarterly*, Vol. 85, No. 3 (Jul-Sep 2014), pp. 14–19, 28.

4. N. S. Fitz and P. B. Reiner, The challenge of crafting policy for do-it-yourself brain stimulation, *Journal of Medical Ethics*, Vol. 41, No. 5 (May 2015), pp. 410–412.

5. Transcranial Direct Current Stimulation, https://www.reddit.com/r/tDCS/.

6. C. H. Wu, Electric fish and the discovery of animal electricity, *American Scientist*, Vol. 72, No. 6 (November-December 1984), pp. 598–607.

7. C. I. Sarmiento et al., Brief history of transcranial direct current stimulation (tDCS): From electric fishes to microcontrollers, *Psychological Medicine*, Vol. 46, No. 15 (November 2016), pp. 3259–3261.

8. M. S. George, Stimulating the brain, *Scientific American*, Vol. 289, No. 3 (September 2003), pp. 66–73.

9. J.-P. L. et al., Evidence-based guidelines on the therapeutic use of transcranial direct current stimulation (tDCS), *Clinical Neurophysiology*, Vol. 128, No. 1 (Jan 2017), pp. 56–92.

10. M. Bikson et al., What psychiatrists need to know about transcranial direct current stimulation, *Psychiatric Times*, Oct. 27, 2017. https://www.psychiatrictimes.com/view/what-psychiatrists-need-know-about-transcranial-direct-current-stimulation.

11. J. C. Horvath et al., Quantitative review finds no evidence of cognitive effects in healthy populations from single-session transcranial direct current stimulation (tDCS), *Brain Stimulation*, Vol. 8, No. 3 (May-Jun 2015): pp. 535–50.

12. J. Nelson et al., The effects of transcranial direct current stimulation (tDCS) on multitasking throughput capacity, *Frontiers in Human Neuroscience,* Vol. 10 (Nov 29, 2016), p. 589.

13. J. Woodruff, How a gentle electrical jolt can focus a sluggish mind, *PBS NewsHour*, Mar 31, 2015, https://www.pbs.org/newshour/show/gentle-electrical-jolt-can-focus-sluggish-mind.

14. A. Wexler and P. B. Reiner, Oversight of direct-to-consumer neurotechnologies, *Science,* Vol. 363, No. 6424 (Jan 18, 2019), pp. 234–235.

15. E. Klein et al., Engineering the brain: Ethical issues and the introduction of neural devices, *The Hastings Center Report*, Vol. 45, No. 6 (November-December 2015), pp. 26–35.

16. S. Perkowitz, Bad blood, worse ethics, *Los Angeles Review of Books*, Sept. 7, 2018. https://lareviewofbooks.org/article/bad-blood-worse-ethics/#!.

17. A. Hartmans and P. Leskin, The rise and fall of Elizabeth Holmes, the Theranos founder whose federal fraud trial is delayed until 2021, *Business Insider*, Aug 11, 2020. https://www.businessinsider.com/theranos-founder-ceo-elizabeth-holmes-life-story-bio-2018-4.

Doing the Math

Introduction

Science writers find it difficult to convey ideas about mathematics, statistics, and risk assessment to general readers. The reasons range from math anxiety, apparently prevalent even among educated people, to the difficulty of translating arcane math language and symbols into ordinary language. This section is about books and films that try to present math understandably.

"A Short Take on Mathematics" (2015) compares the recent book *Math Geek* to two other popular math books and considers how teaching and learning math have changed in the digital age. "The Poetry and Prose of Math: Part 1, Poetry" (2019) and "The Poetry and Prose of Math: Part 2, Prose (2019)," draw on films from the Labocine archive (https://www.labocine.com). "Part 1" features films about abstract math, such as a profile of an artist who transmutes equations into images of objects occupying more than three dimensions. "Part 2" discusses films about applied math, such as methods to predict the spread of disease, and about what happens when basic applied math, arithmetic, is politicized, much like when true facts are distorted and politicized.

Science Sketches: The Universe from Different Angles
Sidney Perkowitz
Copyright © 2022 Jenny Stanford Publishing Pte. Ltd.
ISBN 978-981-4877-94-7 (Hardcover), 978-1-003-27496-4 (eBook)
www.jennystanford.com

"Infinity on Screen" (2016) covers the film *The Man Who Knew Infinity* about the brilliant self-taught Indian mathematician Srinivasa Ramanujan. The film illustrates the inexplicable nature of mathematical genius and how Ramanujan was penalized as an outsider in early 20th century British society and academia, despite his mentoring by Cambridge University mathematician G. H. Hardy. The piece also compares the film to others about math and mathematicians, and shows the kind of abstract math that mathematicians think about.

A Short Take on Mathematics

Reviews of *Math Geek: From Klein Bottles to Chaos Theory, a Guide to the Nerdiest Math Facts, Theorems, and Equations,* Raphael Rosen (Adams Media, 2015); *Mathematics for the Million*; Lancelot Hogben (W. W. Norton, 1993); and *The Joy of x: A Guided Tour of Math, from One to Infinity,* Steven Strogatz (Houghton Mifflin Harcourt, 2012).

I learned a lot from Raphael Rosen's *Math Geek*, not just about math, but about how those who like math are perceived and how to reach those who don't much enjoy it. These are topics we need to consider as we figure out how to better educate Americans about science, and how to get more students and more diverse ones ready for careers in science and technology.

Rosen, who has worked at San Francisco's hands-on Exploratorium science museum and writes about science for various outlets, has produced an unusual math book. For one thing, rather than written in chapters, the book features brief essays about math and its appearances in daily life, which makes it more suitable for dipping into than for sustained reading. For another, *Math Geek* may be the first math book ever to explicitly welcome "geeks" and "nerds."

This made me wonder: those two terms are used a lot, but what do they really mean? Both can claim distinguished lineage: "geek" traces back to Shakespeare and Jack Kerouac, and "nerd" may come from an old Dr. Seuss story. They also share dark undertones, especially "geek," which once meant a circus performer whose act consisted of biting off the head of a live chicken. Now, according to the Oxford English Dictionary, a geek is an "unsociable person obsessively devoted to a particular pursuit" and a nerd is "socially inept" and exclusively dedicated to "an unfashionable or highly technical interest." Neither suggests bizarre circus acts, but they carry similar less-than-complimentary connotations. (You can, of course, find plenty of online discussion about subtle differences between a "geek" and a "nerd.")

Yet there's also some underlying admiration for nerds and geeks in our society. Unsociable and obsessed they may be, but in today's hi-tech world, they can be highly successful too. As one online commenter put it, "A nerd is the guy you made fun of in high school who you work for today." I don't know if anyone ever made fun of entrepreneurs like Bill Gates or Elon Musk when they were teens, but the dedicated focus of technology gurus like them has produced companies that employ lots of people and, besides that, are changing the world.

In addressing *Math Geek* to nerds and geeks, Rosen and his publisher are on to something meaningful that may also sell books. If so, is he preaching only to the already converted who pursue the "unfashionable" area of mathematics? Rosen's answer would be "no," because as his introduction makes clear, his intention is to reach a much larger audience. He wants to convince anyone who reads his book that mathematics is not "just a series of rote exercises performed in a classroom." Instead, it is "built into the fabric of reality [and] is a living feature of the world we live in [...] Math has a beauty that can stop you in your tracks." Rosen's hope is to turn non-nerds and would-be geeks into true appreciators of the mathematics in their lives—and maybe, who knows, that will encourage some of them toward careers in math or science.

Rosen's goal raises another question the book inspired me to explore: how do you get math, or appreciation of its wonders, across to readers who are interested but lack the tools to really get into the subject? Math is expressed in a special, highly symbolic language that must be learned, starting with algebra. That may be one reason why "math anxiety" often gets between people and their engagement with math. It isn't clear that this special kind of anxiety is a well-defined mental state, but anecdotal evidence suggests that some people feel a kind of mind freeze when confronted with a mix of numbers, letters like the dread algebraic unknown "x," and exotic terms like "cosecant" and "differential calculus."

As a result, math is harder to popularize than other categories of science. People are awestruck by the megaphysics of black holes and galaxies and the nanophysics of quantum mechanics and the Higgs boson; space exploration is always exciting; the natural wonders of our Earth, from volcanoes to exotic animals, elicit strong interest; and much of the science in biomedicine, like the current explosion in neuroscience, has the twin draw of illuminating our own nature and producing useful medical outcomes. But even though mathematics is a substrate and a framework for science, its abstractness and its special language make it hard to access and hard to link to the everyday world.

Still, mathematicians and science writers have done their best to popularize math. Looking at their efforts provides lessons in how to get math across and shows how writing about scientific ideas has evolved in the digital age, as exemplified by *Math Geek*.

One early popular math book I tackled as a kid wanting to be a scientist will celebrate its 80th birthday next year. It is Lancelot Hogben's *Mathematics for the Million: How to Master the Magic of Numbers*. In 1936, Hogben, a British zoologist and medical statistician, produced a classic that can still be found today as a paperback for sale on Amazon, where it continues to get admiring reader's comments. Its blurbs on the back cover, old as they are, are hard to beat. One is from Albert Einstein, who says the book "makes alive the contents of the elements of mathematics"; the other is from H. G. Wells, who calls it a "great book." At over 650 pages, it is also a big book, and no wonder. It teaches math through the long history of the subject, from the first insights into numbers and arithmetic by ancient civilizations to the great 17th century developments of calculus and probability theory by Newton, Pascal, and others that underlie much modern mathematics. Well-illustrated with figures and equations, the book even has mathematical exercises, with answers supplied, so the reader can test his or her knowledge. By current standards, it is more a textbook than a popular work, though enlivened by Hogben's discourses on mathematics and mathematicians through the centuries. Hogben himself understood his book as no easy read. His prologue exhorts the reader in capital letters, "WHAT YOU GET OUT OF THE BOOK DEPENDS ON YOUR CO-OPERATION IN THE SOCIAL BUSINESS OF LEARNING." It is a "social" compact between reader and author to work hard to gain understanding.

This contrasts with a recent math book, Steven Strogatz's *The Joy of x: A Guided Tour of Math, from One to Infinity* (2012). Strogatz, a working mathematician at Cornell University who specializes in applied mathematics, extended his series of popular columns about math for *The New York Times* into this book. Like *Mathematics for the Million*, *The Joy of x* begins with the basics of numbers and arithmetic, then follows an ascending arc to higher levels but not necessarily in chronological order. Rather, its chapters appear under headings that give a flavor of the basic areas that math covers: numbers, relations (algebra and beyond), shapes (geometry and mathematical proof), change (calculus), data (probability theory), and frontiers (miscellaneous topics, from prime numbers to Moebius strips and infinity).

Rather than intersperse historical tidbits with mathematical explanations, Strogatz gives vivid examples of math in daily life, tied to today's world and its pop culture. For instance, he writes "The best introduction to numbers I've ever seen [...] appears in a *Sesame Street* video called 123 Count with Me," then uses the video to comment on some of the deeper philosophical meanings of numbers. Strogatz has a light and entertaining touch, and his chapters, typically under 10 pages long, are easy to digest. This makes the book eminently readable in a way that Hogben is not.

That is a plus for a popular book about mathematics. It did, however, make me wonder about the much-touted loss of concentration and inability to do "long reads," supposedly engendered by our obsession with reading short online click-worthy pieces. Is this affecting how we all read and write now, and does writing about science and math need to change to accommodate this newer style of absorbing information?

This trend seems more pronounced in Rosen's book than in Strogatz's. *Math Geek* is similarly organized under major headings—Shapes, Behavior, Patterns, and Special Numbers—but instead of chapters, it has 100 short essays about mathematical topics. Typically a couple of pages long, the essays could work as a screen's worth of reading if they were downloaded from the Internet. Compared even to short, easily digestible chapters, these are bite sized.

Their topics will be familiar to anyone with a math background, but Rosen makes them enticing to beginners by writing in a relaxed conversational style, assuming little math knowledge, and relating the math to pop culture and ordinary events. The essays have intriguing or provocative titles, such as, "Why are manhole covers round?" (because they can't fall through their own openings); "Are you living in the fourth dimension?" about the mathematical oddities called Klein bottles and the shape of the Universe; "Your social media jealousy has mathematical roots," about the result called the friendship paradox that explains why everyone else has more friends than you do; "What the subway map leaves out," an introduction to topology; and "Did you inhale Caesar's last breath?" a statistical analysis of the odds that air molecules that Julius Caesar exhaled when he died still linger in the atmosphere.

Consistent with his strategy of short pieces, Rosen has mostly picked topics in the theory of numbers, geometry, and so on that do not need long exposition. He does cover infinite series and the concept of infinity, but nowhere in the book do you see calculus or very many equations. Illustrations help, and also, to provide more heft, each essay names the mathematical concept that it covers and is followed by a short note that extends the concept or otherwise adds meat to the discussion.

Packaging mathematical ideas into short essays means that they cannot be covered in depth, and the connections among them do not jump out. What makes the book appealing to the casual reader might diminish its value for those inspired to dig deeper. One easy fix would be to add specific outside references to each essay in *Math Geek*. These could be discreetly tucked into the back of the book for those who want to delve further, without making it look like a scholarly rather than a popular work.

This does not mean, though, that I'm advocating for a tome like *Mathematics for the Million* as the only right way to popularize math. It's a good thing to have a spectrum of approaches available. While some readers in the digital age respond best to short pieces, others gravitate to long, detailed discussions. All will become more scientifically literate, and some will go on to become scientists, when they will really have to grapple with dense and serious writing. But it doesn't matter if they get there by first plowing through Hogben, or by feeling that "zing!" of passionate interest through reading one of Rosen's essays—they will be on their way to becoming the best kind of geek or nerd, the kind that our society needs.

The Poetry and Prose of Math: Part 1, Poetry

Reviews of the films *Matter Patterns, The Mathematics Engraver,* and *Logically Policed* (These films can be seen at the Labocine site https://www.labocine.com.)

The 32 films in Labocine's August selection "The Poetry of Math" include both poetic and prosaic mathematical moments, and to carry the literary analogy further, fictionalized moments as well. All these are welcome. People like me who write about science know that showing non-mathematicians and non-scientists what math is all about, from its abstractions to its applications, is a big challenge. Displaying the richness of math in visual and cinematic terms can help people understand the prose, math applied to the sciences and to society; and the poetry, math as an innate human capacity and an adventure of the human spirit that has much in common with art. In this article I'll talk about the poetry and how to convey it. I'll move on to the prose in the second article.

In 1623, the great Renaissance figure Galileo Galilei, a founder of modern science, wrote that the book of nature is "written in the language of mathematics." Math is a product of the human mind, but it has been inspired by nature and we in turn use it to analyze nature. Some think that seeing nature through a mathematical lens diminishes the appreciation of its beauty and wonder; but I think these insights only enhance the appreciation, part of the poetry of math.

That view is shared in the documentary *Matter Patterns* (2014, Olga Yakimenko). Combining a narrator's comments with images of nature, the film illustrates the abstract idea of self-similarity, where the repeated application of simple rules produces much of the world's enormous complexity. As the film's images illustrate, self-similarity determines how trees branch and rivers flow, how color is distributed on butterfly wings and zebras, and how biological cells change and combine. At the end of a stunning visual tour, the narrator restates the film's premise that it's "nonsense" to conclude that seeing nature mathematically "can somehow make beauty evaporate." Instead, we can "develop understanding where there was once only seeing"—a strong argument for the power of mathematics to broaden our perception of the world and its beauty.

Math can also carry us into worlds we have never seen. The documentary *The Mathematics Engraver* (2016, Quentin Lazzarotto) displays this magical ability through the work of an unusual artist. Patrice Jeener lives in the medieval village La Motte-Chalancon in south-eastern France, where for decades he has taken images from mathematics as the inspiration for his art.

Any mathematical equation can be turned into an image by graphing it. Simple equations yield simple flat shapes such as circles. Intricate equations produce other-worldly three-dimensional shapes that curve, undulate, coil, and twist in ways rarely seen in actual natural or artificial objects. Some of these abstract mathematical objects represent shapes from imaginary worlds with four or more spatial dimensions, projected onto our 3D world. Jeener generates these fantastical shapes from equations and makes etchings of those he finds "pleasing." He copies the image onto a copper sheet and gouges out its curves by small, meticulous bites of a sharp chisel. Then mounting the inked plate in a hand press, he prints replicas on paper until he is satisfied with the quality of the reproduction as the final print.

These visions of abstract math, shown in loving detail throughout the film, evoke varied reactions from mathematicians. One says that the prints are like special glasses "for seeing things that we couldn't even imagine being able to construct or touch." Another notes that the shapes Jeener finds artistically elegant are also mathematically elegant, in that they represent the most compact, efficient solutions to certain problems. Jeener shares another trait with mathematicians, who "enjoy getting away from reality" as one of them says. Jeener too escapes reality in his work. "I'd like to have lived in a space I'd thought up myself," he says, as he imagines himself walking there. Mathematicians enter that world too, but it is Jeener who brings back pictures of his travels.

How and why does anyone enter that abstract world sufficiently to appreciate the poetry of math or create new poetry as a mathematician? Why do some people naturally take to math with pleasure and even pursue it professionally, whereas many others fear it? *Logically Policed* (2014, Damiano Petrucci) offers clues to answers and to the origins of mathematical abstraction.

Interviewing mathematicians and math communicators, the film elicits their childhood experiences with numbers that inspired them, such as mentally making change in a family shop within the complicated English system of pounds, shillings, and pence; or being challenged to use speed and distance to find arrival times for trips in the family car. Sara Santos, founder of MathsBusking, which provides entertaining mathematical street experiences, explains how such encounters lead to higher levels. The first person to look at two stones, two sheep, and two grapes, she says, and realize that these disparate objects share "twoness," was making a leap into mathematical abstraction. Because math is not a collection of equations but a mental activity, she adds, evolution has since then carried us to our present levels of math and its abstractions.

The power of abstraction to encompass a multitude of things gives pure math its real-world influence in telecommunications, laser technology, smartphones, and more, as the interviewees point out. But learning these abstractions requires good math teaching, which is in short supply, and persistence from the student. To become a creative mathematician, it also requires what one interviewee calls "a thorn in the side," the unbearable mental itch that makes a person relentlessly seek the answer to a math problem. We never know for sure that the answer, once found, will spark important new mathematical thought or lead to a significant real-world result. But the interviewees present examples where exercises in pure math, such as proposing numbers involving the square root of -1, have had profound real outcomes.

That is where the poetry turns into prose, as I'll discuss in Part 2.

The Poetry and Prose of Math: Part 2, Prose

Reviews of the films *Einstein–Rosen*, *Strange Particles*, *Persistence & Vision*, and *Alternative Math* (These films can be seen at the Labocine site https://www.labocine.com.)

The films in Labocine's August selection "The Poetry of Math" actually included both poetic and prosaic mathematical moments, which I equated to pure and applied math, respectively. I covered poetry in the first of two articles, and now I'm turning to the prose of applied math.

What do we mean by applied math? In one sense, it's everywhere. Every artifact around us was designed and built with the help of numerical measurements and calculations, whether simple or advanced. Basic arithmetic and geometry show up as well in daily activities such as calculating the tip for a restaurant meal or finding the area of a room. At this level, math is so woven into society that we take it and its basic ideas for granted.

Apart from its own standing as a special area of knowledge and research, math is also an essential scientific tool. Science could not function without the quantitative and analytical capabilities math provides, combined with observation and experiment. This connection is strongest for the physical sciences and engineering, especially physics, but is also important in the social sciences such as economics and sociology, typically through statistical analysis. Mathematics underpins computer science and plays a growing role in biomedical science as that becomes more analytical. Several films in "The Poetry of Math" illustrate both kinds of use.

Einstein–Rosen (2017, Olga Osorio) shows math applied to physics through its story of two young brothers, Teo and Oscar, and a wormhole. After Einstein developed the equations of general relativity, his theory of gravity, he and others explored their solutions. Two results were especially intriguing: cosmic locations where gravity becomes so strong that nothing can escape, called black holes; and "Einstein–Rosen bridges," later called wormholes, short cuts between different regions that might be very far apart in space or time, perhaps making it possible to travel rapidly between them. Both were purely mathematical predictions with no guarantee that they actually exist; but astronomers have found clear evidence of black holes, including one at the center of our own galaxy. They are now thought to be real objects.

No wormhole has ever been observed, but in *Einstein–Rosen*, Teo believes he has found one. Oscar doubts it, but Teo kicks a soccer ball toward the supposed wormhole, and...the ball disappears. Now fast forward from 1982, when Teo found the wormhole, to 2017 with the brothers now in their forties. Teo has calculated that the ball will travel 35 years forward in time to this exact moment. Sure enough, it reappears as the brothers watch. Then to their surprise, dozens of other objects rain down, mainly Oscar's toys that Teo has mischievously tossed into the wormhole. The last straw is the appearance of Oscar's pet turtle Pepe, who Teo had also catapulted into the wormhole. Oscar is upset, saying "I cried for six months, thinking he had been run over by a car;" but at least Pepe has returned, apparently unharmed by his trip through time.

The nearly feature-length *Strange Particles* (2018, Denis Klebleev) shows the influence of math on physics in a different way. Konstantin is a 30-ish theoretical physicist, who does not do lab experiments but works with complicated equations (briefly shown in the film) describing quantum physics and black holes. He teaches university physics in Moscow. In the summer, he teaches and is a dorm counselor at the school's sea-side campus, where the film is mostly set.

We first meet him walking along a road, talking on his cell phone about his research. He is devoted to his work, but it isolates him as he spends most of his time thinking about it. He makes no friends at the summer campus and eats alone. The intensity carries over to his teaching. His students are after all young men who may be willing to learn during the summer but are also drawn to summer pleasures, soccer, and young women. Konstantin cannot accept their lack of commitment and constantly harangues them to work harder, which only makes them sullen. He too is frustrated and unhappy; we hardly see him smile or laugh during the whole film.

One chance at enlightenment comes from a student, who says he quit striving when his early promise in chess was eclipsed by a younger but more brilliant player. He adds, "we must be realistic...I am not a genius. That's it...it's the same in science. What's the point of being a scientist if you are not a genius?" This seems to impress Konstantin, but at the film's end, we do not know his future in physics and in life. The film's write-up describes Konstantin as living in a quantum world that does not mesh with the classical one. I don't think that quantum physics is the issue. I see Konstantin rather as choosing to live with abstractions remote from ordinary life, but unable to make this satisfying. In contrast, Patrice Jeener in *The Mathematics Engraver* also lives with mathematical abstraction but says "I'm happier all the time about doing the job I do."

Mathematical thinking has been less dominant in biomedicine than in physics, but it is important in epidemiology to study how infectious diseases spread. *Persistence & Vision* (2017, Mahalia Lepage and Jeremiah Yarmie) gives a brief but effective introduction to the subject and the application of math to a problem with definite real-world, even life-and-death, consequences.

The film is narrated by Ryan Sherbo, an earnest looking undergraduate who won an award for his epidemiology research at the University of Manitoba. Accompanied by animations that enhance his comments, he explains how "biomath" can predict how quickly tuberculosis will spread in households. Like all true scientists and scientists-in-training, he feels the emotional rush when the research comes together and you "really feel like you've accomplished something." He moves beyond the science as well to give an important reminder that math-based research (and any other kind) has impact beyond the lab: "The most inspiring thing... is the idea that this can really be useful. It's really about...making sure your research is easy to interpret [so that] nothing gets lost in translation from the math side to the biology side to the social science side to the political side."

Research can indeed have political ramifications, and math itself can be politicized in the public arena through manipulated statistics and rigged elections. *Alternative Math* (2017, David Maddox) chillingly comments on our time of "fake news" and "alternative facts" through its wickedly entertaining tale about what happens when the bedrock truths of math come under fire.

It starts when math teacher Mrs. Wells faces disgruntled student Danny who earned an "F" on his addition test. He thinks $2 + 2 = 22$, not 4, and won't back down even when Mrs. Wells tries to set him right. Danny's parents show up, call Mrs. Wells a Nazi, a tramp, and a bitch for saying her answer is right and Danny's is wrong, and slap her. The school principal supports them, saying "parents don't want you ramming your biased views down their kid's throats." Crowds demonstrate carrying signs like "God hates facts," while the media call the flap MathGate and interview experts to debate whether the correct answer is 4 or 22. Finally, Mrs. Wells is called in to be fired, but she has the last laugh. When the principal offers her two severance checks of $2000 each for a total of $4000, she smiles and responds "Wrong! It's $22,000."

The seven films I've reviewed in Parts 1 and 2 are only a fraction of the 32 presented in "The Poetry of Math." A look at the others will further enlarge anyone's understanding of math and mathematicians.

Infinity on Screen

Reviews of the films *Jurassic Park* (1993), *Good Will Hunting* (1997), *Straw Dogs* (1971), *Proof* (2005), *Pi* (1998), *A Beautiful Mind* (2001), *The Imitation Game* (2014), *The Theory of Everything* (2014), and *The Man Who Knew Infinity* (2015).

In 1914, Srinivasa Ramanujan, a 26-year-old clerk and an aspiring mathematician, arrived at Cambridge University from Madras, India. He had sent his ideas to the Cambridge mathematician G. H. Hardy, whose background was very different from Ramanujan's. Hardy was a product of upper-level British education, an honors graduate of Cambridge's Trinity College who became eminent during his subsequent career at Cambridge and Oxford. Ramanujan, born to a clerk and a housewife in south India, had no such benefits. He was largely self-taught without a college degree, yet what he sent to Hardy revealed novel and significant mathematical results. Hardy, himself a former child prodigy in math, recognized Ramanujan's innate genius and invited him to Cambridge.

Their resulting mentor–student relationship and collaboration helped Ramanujan produce important results in areas of pure mathematics such as number theory, the study of the positive integers. It also prompted a newly released film about him, *The Man Who Knew Infinity*. Written and directed by Matt Brown and based on the book of the same name by Robert Kanigel, it stars Dev Patel (recently seen in *The Second Best Exotic Marigold Hotel* (2015)) as Ramanujan and the Oscar-winning actor Jeremy Irons as Hardy.

The Man Who Knew Infinity takes its place among films about a topic that is hard to express on screen, mathematics—especially pure math that studies abstract mathematical ideas without any applications in mind. Science-based stories about space exploration, natural disasters or biotechnology can show something tangible and engaging like a spaceship launch or a cloned creature. But though mathematics underlies all science and technology, it exists mostly in the minds of mathematicians and as symbols written on blackboards or formulas residing in computers. How can this be turned into a film that, like *Infinity*, is meant to tell a dramatic story?

The usual approach is to focus on mathematicians, not mathematics. Like Ramanujan, some are real; others are fictional characters who supply math wisdom or represent certain types. In *Jurassic Park* (1993), a chaos theorist played by Jeff Goldblum worries about the impact of creating living dinosaurs through DNA technology. The "butterfly effect," he explains, shows that small changes can produce large outcomes; or more simply, we should expect unpredictability. Sure enough, the dinosaurs wreak unforeseen havoc before they are brought under control. In *Good Will Hunting* (1997), Matt Damon is a young mathematical genius working as a janitor at MIT. He is mentored by a stereotypical math professor but cannot develop his talent until he overcomes his humble background and emotional damage from childhood abuse. The story has parallels to Ramanujan's, which is mentioned in the movie.

Other characters are identified as mathematicians, but we wonder why, because they do precious little math. In *Straw Dogs* (1971), David Sumner (Dustin Hoffman) is so committed to the mathematics of astrophysics that he neglects his wife Amy. She turns to a former lover, who becomes involved in her violent rape. This and other hostile acts from a gang of hoodlums change David from a disengaged intellectual to a primal creature who ferociously defends himself and Amy. They take brutal revenge on the rapists, ending a film that has nothing to do with math and could have told the same story with David as a writer or artist.

In *Proof* (2005), from the stage play of the same name, Gwyneth Paltrow is Catherine, who gives up dreams of mathematical glory to care for her insane father Robert, once a brilliant mathematician. When Robert dies, his graduate student finds a major new mathematical proof apparently derived by Robert, but Catherine insists it is her work. However, we never see anything about the proof to convince us of its significance. The film's themes of obsession, ambition, and links between creative intensity and madness could just as well have been expressed in a story about a family blessed and cursed with musical genius.

True mathematical madness is central to *Pi* (1998), directed by Darren Aronofsky. In it, Max Cohen (Sean Gullette) is a number theorist who thinks that "Mathematics is the language of nature. Everything around us can be represented and understood through numbers." He finds a particular 216-digit number that he believes opens doors from stock market predictions to the true name of God within Jewish mysticism. When his behavior becomes so erratic that his old mentor says "This is insanity," Max replies "Or maybe it's genius." Whichever, he develops violent headaches and operates on his own brain with a power drill. He survives, and in a final scene sits quietly on a park bench, with his mathematical talent gone but no longer overwhelmed by dark mathematical thoughts.

The films that best compare to *The Man Who Knew Infinity*, however, are about real mathematicians. The most successful of these cinematically and mathematically is *A Beautiful Mind* (2001) with four Academy Awards, including Best Picture and Best Director (Ron Howard), for its story about Princeton University mathematician John Nash, played by Russell Crowe.

Like *Proof* and *Pi*, this film explores madness, but real, not fictional. Nash was awarded the 1994 Nobel Prize in Economic Sciences for his research with important implications in economics and game theory, but he reached this pinnacle only after a grueling personal history. Diagnosed with paranoid schizophrenia in the late 1950s, he spent years in mental hospitals before experiencing a slow recovery and finally returning to research in the mid-1980s.

The film brings viewers inside Nash's mind as he sinks into madness and then haltingly returns to sanity, and also illustrates one of his achievements, the "Nash equilibrium." This is the insight that the best overall outcome for a group of competitors comes if each chooses a strategy assuming that the others also make decisions reflecting the choices of all of them. In an entertaining scene, Nash and his fellow male graduate students mingle with a group of young women in a bar. Nash tells the other men that if each chats up only the prettiest woman, their opposing efforts will add up to nothing; but if each instead talks to a different one of the other women, they're all more likely to make contact with the maximum good outcome for all—a persuasive example of the Nash equilibrium in action.

Two other math-based biofilms were released in 2014. *The Imitation Game* is about British mathematician Alan Turing (Benedict Cumberbatch), a founder of computer science who helped break the German Enigma code in World War II. His life as a gay man when homosexual acts were illegal in the U. K. is a big part of the movie. In *The Theory of Everything*, Stephen Hawking (Eddy Redmayne, who won the 2015 Oscar for Best Actor) is the mathematical physicist who is confined to a wheelchair with a nerve disease but who overcomes this limitation. Each film alludes to the work done by its protagonist, but neither film has a moment of mathematical clarity like the Nash equilibrium scene in *A Beautiful Mind*.

Like Nash, Turing, and Hawking, in *The Man Who Knew Infinity*, Ramanujan has great native ability but encounters great difficulties. He has an innate grasp of numbers and mathematical concepts, shown in the film when as a clerk in India he calculates in his head without an abacus and when his wife says that he is known to "love numbers more than people."

But lacking formal training, he does not realize that sheer brilliance is not enough. He sees the solution to a problem through some higher intuition that he believes comes from God, but does not know how to trace the steps from posing a problem to arriving at the answer, that is, constructing a proof to show other mathematicians that the answer is correct. Hardy teaches Ramanujan to do this. He also carries out proofs of some of Ramanujan's results for publication under his student's name, to Ramanujan's great pleasure.

Figure 1 In the biographical film *The Man Who Knew Infinity* (2015), mathematical genius Srinivasa Ramanujan (Dev Patel) pays homage to Isaac Newton after arriving in England from India. Despite support from his mentor at Cambridge University, he encounters bias that affects his work and life.

However, as an Indian entering English culture, Ramanujan also has special problems. Hardy and others are kind to him as he adapts to life at Cambridge. But at a time when Indians are a subject race within the British Empire, some faculty are shown as condescending or racist. Ramanujan even encounters violence according to one scene in the film. Other issues are his unfamiliarity with Cambridge's arcane rules such as who is allowed to walk on the grass in the quad, and his difficulties in finding the vegetarian meals his religion requires.

Adding to his unhappiness, he has left his wife behind. (In reality, he had been placed in an arranged marriage to a very young girl, but the film makes her nearer his own age and cuts back and forth from his life at Cambridge to hers in India). Most seriously, Ramanujan suffers from poor health for much of his life, culminating in tuberculosis that he develops in England, perhaps because he is unused to the cold climate. This kills him in 1920 at the age of 32.

In his brief time, Ramanujan makes astonishing progress at Cambridge. He becomes a Fellow of Trinity College and is elected to the Royal Society, only the second Indian to receive this honor. The film shows something of his work and its value as Hardy reflects on the great mathematical legacy left by Trinity College's own Isaac Newton, and other eminences like the 18th century Swiss Leonhard Euler and the 19th century German Carl Gustav Jacobi. Both worked in number theory like Ramanujan, who Hardy puts firmly in their exalted company.

Since number theory studies the ordinary counting numbers 1, 2, 3, ..., it may seem simple, but it is a fundamental area that raises many difficult questions. The film makes an effort to show one of these abstract issues as it explains about partitions, the number of different ways that an integer can be written as a sum of integers. For example, 4 can be written in five different ways: 4, 3 + 1, 2 + 2, 2 + 1 + 1, and 1 + 1 + 1 + 1. The partition value climbs rapidly: 20 can be partitioned in 627 ways, and 100 in 190,569,292 ways. Number theorists had long sought a way to calculate the partition value for any arbitrary integer without laboriously counting up all the possibilities. As the film shows, Ramanujan and Hardy find a partial answer in 1918 with a famous breakthrough, a formula that gives the approximate partition number for any integer and becomes more accurate the bigger the integer.

Compared to the applied math carried out by Nash, Turing, and Hawking, one can reasonably ask, what's the value of such pure mathematics? Surprisingly, number theory has links with quantum physics and the quantum properties of black holes. This hints at intriguing but so far inexplicable connections between the abstract idea of numbers and the real physical world. Had Ramanujan lived into the era when quantum physics was being developed, his marvelous intuition might have revealed these tantalizing possibilities decades earlier. But his insights still play a role through the notebooks he left, which researchers continue to investigate.

The Man Who Knew Infinity evokes human interest by showing the tension between Ramanujan's phenomenal grasp of mathematics and his struggles to fully use his gift. The film displays moments of mathematical exposition as well. However, the drama is not fully developed. The scenes with Ramanujan's wife are unconvincing, and his own circumstances limit what we can know and feel about him. Like many natural geniuses, he can say little about how his ideas arise, and his early death precludes following his development as man and mathematician.

But the film tells another story as well, about Ramanujan's mentor Hardy. His life represents a future path that Ramanujan himself might have taken: a young prodigy growing into a man who says, as Hardy does in the film, "mathematics is my church" and devotes himself to it. Within this commitment, Hardy becomes Ramanujan's friend and promoter, advocating for the honors he deserves in the face of anti-Indian sentiment.

Ramanujan's brief shining mathematical arc is the main theme of the film, but Hardy's longer history of dedication to mathematics and his support for Ramanujan deepen the story. This enhances the film as one of the few about real mathematics and not one but two real mathematicians. Perhaps a better title would have been "The Men Who Knew Infinity."

Four Scientists and a Quest

Introduction

The story of science is also the story of scientists. We have already met the pure mathematicians Srinivasa Ramanujan and G. H. Hardy in "Infinity on Screen." Here I consider experimental scientists, starting with the great 17th century figure Galileo Galilei in "Galileo Through a Lens: Views of His Life and Work on Stage and Screen" (2009). He stands out for his telescopic observations that changed views of humanity in the cosmos. Less well known, despite many works on stage and in the media about him, is that Galileo used a similar arrangement of lenses as a microscope, enhancing his stature as a scientist whose curiosity led him to study nature at all scales.

"The Shadow of Enlightenment" (2010) reviews the book of that name by historian of science Theresa Levitt, about the early 19th century French optical scientists François Arago and Jean-Baptiste Biot who studied the polarization of light and the theory of colors. After the French Revolution, their political views were oppositely polarized in a different way, which affected their approaches to color theory. Research rivalry further harmed their collaboration. This historical story suggests that personal rivalry and the mix of science and politics have been and probably always will be part of science.

Science Sketches: The Universe from Different Angles
Sidney Perkowitz
Copyright © 2022 Jenny Stanford Publishing Pte. Ltd.
ISBN 978-981-4877-94-7 (Hardcover), 978-1-003-27496-4 (eBook)
www.jennystanford.com

"If Only 19th Century America Had Listened to a Woman Scientist" (2019) presents the mid-19th century research of Eunice Foote, first to discover the major roles of carbon dioxide and water vapor in climate change. Her results did not receive wide attention or proper credit until recently, because Foote was marginalized as a part-time and female scientist. This is just one indication of what society loses by excluding certain groups from science.

"Romancing the Quantum" (1991) is not about a specific scientist, but about what drives many scientists, a kind of romantic quest in their search for truth. This does not, however, mean that only pure science deserves support, such as was requested for the Superconducting Super Collider in the era when I wrote the piece. Science that contributes to humanity must be weighed too, as we have seen with the COVID-19 pandemic.

Galileo Through a Lens: Views of His Life and Work on Stage and Screen

When Galileo Galilei first turned his telescopic lenses toward the skies in 1609, he set in motion a train of events that would lead the world to examine him and his science through its own lenses. That examination continues today. One way to understand what it reveals is to compare what Galileo actually did to how his life and work appear in dramatic works for stage and screen.

Galileo's astronomical observations are well known. Among them, he discovered the four largest moons of Jupiter, saw the rough cratered surface of the Moon and spots on the Sun, saw (in blurred form) the rings of Saturn, and most important, observed that Venus displays phases, strong evidence for a heliocentric rather than an Earth-centered solar system. That last observation led to Galileo's conflict with the received position of the church in favor of a geocentric system.

These achievements alone would ensure Galileo's scientific importance, but he did something else that is less widely appreciated; he also apparently realized that a telescope and a microscope are not very different—each is an arrangement of two curved pieces of glass in a tube—and made microscopic observations or encouraged them to be made by others.

There are no extant images of anything Galileo saw through a microscope, but several pieces of evidence indicate that he used the instrument. His student John Wedderburn wrote that Galileo was microscopically examining certain "minute animals" and "insects" as early as 1610. Later, in 1623, Galileo alluded to "a Telescope adjusted to see objects very close up" in his book *The Assayer.* In 1624, Giovanni Faber, a member of the new scientific institution the Lincean Academy, wrote of Galileo's demonstration of a fly magnified by his optical device, and Galileo sent the Academy's founder a microscope with complete instructions on how to "see the tiniest things close up."

Few scientists have the opportunity to look at the universe at both large and small scales, nor to do work that strongly affects society and religion. It's not surprising, therefore, that the world has turned its lenses on Galileo and his science by examining him in works for the theater, film, video, and television.

Table 1 lists stage works about Galileo. The best known is Bertolt Brecht's *Life of Galileo,* considered the first play to portray a real scientist in a historical situation. It's significant that the original play in German preceded the dropping of the atomic bomb on Hiroshima in 1945, whereas the American version came after that event; correspondingly, Brecht portrays Galileo as a hero in the earlier version, but as an anti-hero in the later one as Brecht considers the social responsibility of scientists. Among the remaining works on the list, Tom Stoppard and Philip Glass are noteworthy contributors, showing the enduring power of Galileo's story even into the 21st century.

Table 1   Galileo through a dramatic lens: Stage works

Life of Galileo, Bertolt Brecht (German version, 1943; American version, 1947).
Galileo, Tom Stoppard (1970, unpublished)*
The Genius, Howard Brenton (1982)
Space, Tina Landau (1998)
Star Messengers (opera), Paul Zimet and Ellen Meadow (2001)
Galileo Galilei (chamber opera), Philip Glass et al. (2002)**
Galileo's Daughter, Timberlake Wertenbaker (2004)
Galileo Walking Among the Stars, Em Lavik (2004)

---

*A response to Brecht's play
**Based on Brecht's play

Table 2 lists works about Galileo or that use his name in film, video, and television. It's remarkable that the film history dates back to 1909 (though the title for that effort *Galileo, Inventor of the Pendulum* is misleading; Galileo merely observed that the period of a pendulum is independent of its amplitude if the amplitude is not too large). Several of these works are film adaptations of Brecht's play, including the best known one, Joseph Losey's 1975 version of the stage play he had directed earlier. Others take Galileo or his science in unexpected directions. The Japanese TV series *Garireo* (2007), for instance, is about a detective whose cleverness earns him that nickname in honor of the scientist.

Table 2   Galileo through a cinematic lens: Film, video, and television works (entries not otherwise described are films for theatrical release)

Galileo, Inventor of the Pendulum (1909)
Galileo (1911)
Leben des Galilei (1947) aka *Galileo**
The Life of Galileo (1962 TV movie)*
Galileo (1969)
Galileo (1974)
Galileo (1975, Joseph Losey)*
Topo Galileo (1987)
Galileo Galilei (1989 TV movie)
Galileo (1994)
Galileo (1997 video short)
Galileo Smith Visits the Solar System (1997)
Galileo: On the Shoulders of Giants (1998 TV movie)
Galileo (1998 TV series documentary)
Baby Einstein: Baby Galileo Discovering the Sky (2003 video short)
Garireo (2007 TV series) aka *Galileo*
Galileo Mystery (2007 TV series documentary)
Dear Galileo (2009)
Motel Galileo (2011)

---

*Based on Brecht's play

    Baby Galileo Discovering the Sky (2003) is part of the *Baby Einstein* series meant for very young children. This particular video introduces babies to the universe using "twinkling stars and colorful planets" along with colors and music. *Galileo: On the Shoulders of Giants* (1998) gives Galileo a fictional young apprentice, providing a story about Galileo's science and his troubles with the Church suitable for children aged seven and up.

    Dear Galileo (2009) is an especially interesting semi-metaphorical use of Galileo's story. In this light-hearted Thai production, two college-age girls leave their native Thailand for Europe, one to heal a broken heart and the other to recover from an academic setback. Inspired by Galileo, they vow to visit his birthplace of Pisa. They also promise to remain together no matter what their differences, just as the two different objects that Galileo dropped from the Leaning Tower stayed together as they fell at the same rate.

Viewing Galileo through these lenses gives a clear picture of his legacy in general, and in popular culture, after 400 years. The world well understands that he looked through telescopes, put the Sun in the center of the solar system, and got into trouble for it. He represents the scientist who seeks and states the truth even when it's difficult to do so. Remarkably, his image as scientist and truth seeker is still considered powerful enough to inspire the young. However, the world seems unaware that Galileo also looked through a microscope, but that did not produce the drama of his astronomical findings.

Perhaps the most intriguing outcome, as shown by the film *Dear Galileo,* is that the world "knows" that Galileo dropped things off the Leaning Tower of Pisa. Although Galileo wrote of the experiment, historians agree that in all probability he never actually carried it out. Like Isaac Newton and the falling apple, this story has a mythical power that rivals the science itself, and that's not a bad thing if it helps fix Galileo's true achievements in people's minds.

References

Freedberg, David. *The Eye of the Lynx* (University of Chicago, Chicago, 2002).

Internet Movie Data Base (IMDB), http://www.imdb.com.

Istituto e Museo di Storia della Scienza. *Galileo's Microscope Anthology* http://brunelleschi.imss.fi.it/esplora/microscopio/dswmedia/risorse/anthology.pdf.

Rice University. *The Galileo Project,* http://galileo.rice.edu/

Shepherd-Barr, Kirsten. *Science on Stage: From Doctor Faustus to Copenhagen* (Princeton University Press, Princeton, 2012).

The Shadow of Enlightenment

Review of *The Shadow of Enlightenment: Optical and Political Transparency in France 1789–1848,* Theresa Levitt (Oxford University Press, 2009).

Beyond the political and social disruption caused by the French Revolution, this stupendous event affected much of French and European culture, including science, sometimes in roundabout ways. For instance, the École Central des Travaux Publics—later the École Polytechnique—was founded in 1793 to train engineers who would serve the new revolutionary order. And it was at the Polytechnique that two young men, François Arago and Jean-Baptiste Biot, met, studied under Pierre-Simon Laplace, and went on to become leading optical scientists of the era—and also participants in that new post-revolutionary world.

Although their names are now mostly absent from textbooks, Arago and Biot made many contributions, including considerable work with polarized light. That topic is currently taught in routine ho-hum fashion, but at the time, much about the phenomenon was excitingly new. This is a main thread in Theresa Levitt's absorbing *The Shadow of Enlightenment.* Leavitt, a historian at the University of Mississippi, draws on her background in physics and history of science to unpack scientific, industrial, and cultural implications of "polarization" in the early 19th century. She does the same for other areas in which Arago and Biot worked, such as photography. In what she calls an onion-like scheme, she layers aspects of their science and their linked careers into a grander vision of politics and society during that time of change.

Change there was, but *The Shadow of Enlightenment* also supports the adage "plus ça change, le plus c'est la même chose." Arago and Biot would not recognize modern optics, yet any modern scientist can see familiar elements in their careers and in the surrounding scientific culture: the joy of discovery; the pleasure of collaboration, later spoiled by rivalry and attachment to competing ideas; the interweaving of scientific careers with administrative and political ones; the importance of scientific societies; the economic impact of applied science.

The first collaboration between the two researchers came directly from revolutionary thinking, which ordained a rational system of measurement instead of one based on the size of the king's foot. The new unit, the meter, was defined as the ten-millionth part of a quarter of the Earth's circumference. Starting in 1806, Arago and Biot worked together in Spain to refine earlier erroneous measurements of the circumference. Returning to France (Arago after surviving accusations of spying in Spain), they continued their connection as elected members of the Académie des Sciences and in other ways.

But their relationship deteriorated when they studied polarization, which had been discovered in doubly refracting Iceland Spar (calcite) in the late 17th century. Then, in 1809, Étienne-Louis Malus noticed that light reflected from the windows of the Luxembourg Palace was polarized. Arago quickly built a polarimeter, which used mirrors to create polarized light. When he inserted a thin sheet of mica, he saw what we would now call interference colors—colors so vivid and striking that his displays were popular amusements in the salons of Paris.

Biot soon competed with Arago by constructing his own polarimeter. This led to ill feeling, but there were even deeper differences between the two. According to Leavitt, the disagreement was not the old one of particles *versus* waves; she writes that "neither Arago or Biot cared very much whether light was a particle or a wave." Rather, they quarreled over how to describe the colors their polarimeters created.

This was a broader question than it might seem because of the long-standing inquiry over why objects in the world display the colors they do. In his *Opticks,* Isaac Newton had related the color of an object to its surface properties and constructed an analytical approach based on thin films. Examining the colored rings seen when a curved and a flat piece of glass are pressed together under white light ("Newton's rings"), he related the colors to the thickness of the small air gaps producing them. This analysis was generally accepted, but by the early 19th century, not the idea that it explained the colors of natural bodies.

With their work on polarization, Arago and Biot entered the debate. Adding another layer of meaning, Leavitt notes that their scientific controversy also reflected Biot's deep political conservatism *versus* Arago's strong belief in radical ideals. Biot asserted that Newton's analysis was essential to describe the colors produced by the polarimeter; hence external authority was needed to synchronize what people saw. For Arago, critiquing Newton, "it sufficed to look"; that is, all who saw the color display could see and compare the same colors. (This is reminiscent of reactions after Albert Einstein proposed general relativity, when some felt it was "undemocratic" to support a theory that the average person could not grasp.) As Leavitt puts it, Arago "embraced the forms of civic sociability available in France" through his displays in salons, whereas Biot "saw in Paris society a swirling mess of danger and instability." He spent time away from Paris, returning only in 1821 when Charles X seemed to offer the possibility of an orderly society.

Other strata in Leavitt's explication of "polarization" include what color standards meant to artists of the time and to the flourishing dye industry; and what the study of polarization meant for France's Caribbean colonies that produced sugar, then in huge demand in Europe. In 1815, Biot had begun examining the polarizing properties of sugar in solution as part of research on living *versus* nonliving matter. By mid-century, his technique had become central to the sugar industry as the standard way to characterize the purity and quality of its product.

That industry, however, carried complex political overtones, as the sugar-producing colonies used slave labor. When the French abolitionist movement began gathering strength in the 1830s and 1840s, Arago and Biot took different positions that reflected their philosophies. Biot favored abolition only after the slaves were converted to Christianity; in Leavitt's interpretation, yet another appeal to strong central authority. Arago, who had by this time become heavily involved in radical politics and rose to high positions in France's provisional government of 1848, saw abolition differently; it was a natural extension of human rights as expressed in the original revolution.

Leavitt's study extends to other areas where Arago and Biot contributed, such as astronomy (Arago directed the Paris Observatory for years) and the new science of photography. In summarizing their lives, she broadly characterizes Arago as trusting in political transparency, whereas Biot did not believe that simple, transparent communication could give important answers. This is a useful distinction, especially when recent history in the U. S. and elsewhere shows the continuing importance of political transparency.

However, Leavitt's final comment about the two, "Transparency connected the political and optical, not as a mere metaphor, but ultimately as mechanism," leaves me unconvinced. I do not think the author has demonstrated any linkage beyond the metaphorical between the optical concept of transparency and the political ideal. Indeed, earlier in the book she writes about Arago and Biot: "The goal is not to reduce their scientific beliefs to their politics or vice versa ... no single substratum ... serves as an explanatory key for their actions." Yet, even if the metaphor is only a metaphor, that hardly detracts from Leavitt's scholarly yet imaginative and stimulating reading of two lives in science that illuminates their time and our own.

If Only 19th-Century America Had Listened to a Woman Scientist

Human-induced climate change may seem a purely modern phenomenon. Even in ancient Greece, however, people understood that human activities can change climate. Later, the early U. S. was a lab for observing this as its settlers altered nature. By 1800 it was known that the mass clearing of forests raised temperatures in the Eastern U. S. and that climatic changes followed the pioneers as they spread west.

The causes for such changes, and the understanding that they could have global scope, came from eminent European scientists. Yet an amateur 19th-century American researcher, a woman named Eunice Foote, made a first crucial discovery about global climate change. Her story gives insight into early American science, women in science, and how the understanding of climate has changed. It also reveals how that understanding might have evolved differently to better deal with today's climate problems.

"Foote went from a farmer's daughter to one of the greats in the science of climate change," says John Perlin, a visiting scholar at the University of California, Santa Barbara, and author of books on solar power and other topics. Perlin is currently writing a book about Foote. His extensive archival research and a visit to the locales where she lived and worked have yielded new understanding of her environment and development.

The story of climate change begins in the 1820s, when the French scientist Joseph Fourier discovered the greenhouse effect by recognizing that atmospheric gases must trap the Sun's heat. In 1859 the Irish physicist John Tyndall identified atmospheric water vapor and carbon dioxide gas, CO_2, as main components in absorbing thermal radiation. But in a lost piece of scientific history recovered only in 2011, Foote scooped Tyndall by 3 years when her experiments in 1856 first revealed the roles of water vapor and CO_2. Atmospheric CO_2 levels at the time were only about 290 parts per million (ppm) and global climate change was not yet a known issue. Nevertheless, Foote predicted that changing CO_2 levels could change global temperatures, as we see today with CO_2 at over 400 ppm.

Who was Eunice Foote? At the time of her discovery, she was a 37-year-old wife and mother. Born Eunice Newton in 1819 on a farm in Connecticut and raised largely by an older sister, she lived much of her life in upper New York state—a life marked by events that contributed to her scientific success.

One factor was her excellent schooling. From age 17 to 19, Foote attended the Troy Female Seminary in Troy, New York. This "mecca for women" had been founded by feminist Emma Willard in 1824 as the first women's prep school. It shared facilities with the nearby Rensselaer School (later Rensselaer Polytechnic Institute), whose cofounder Amos Eaton was a particular proponent of hands-on science training. Along with science-heavy curricula, according to Perlin, the schools featured the only two chemistry labs in the world then meant just for students. It was here, he says, that Foote learned lab technique and how to frame and carry out a research project.

Foote was also influenced by a famous neighbor, Elizabeth Cady Stanton, a major figure in early women's rights and suffrage movements who later worked with Susan B. Anthony. Stanton co-organized the first woman's rights convention held in the U. S., which met in Seneca Falls, NY, in 1848. Foote attended and signed the convention's "Declaration of Sentiments" that stated the societal changes necessary to fully include women. More than that, she helped prepare the conference proceedings. Perlin believes she was especially motivated by a resolution that all the professions should be open to women.

As Perlin points out, next to Foote's signature on the Declaration of Sentiments you would have seen her husband Elisha's signature. He was one of only a few male attendees who actually signed, a demonstration of his support that must have been meaningful to Eunice. Elisha also shared her scientific interests. He knew Joseph Henry, the leading American scientist of the time and the founding secretary of the Smithsonian Institution. Elisha carried out meteorological work for the Smithsonian and together with Eunice read Henry's writings about climate.

All these influences contributed to Foote's breakthrough paper, "Circumstances Affecting the Heat of the Sun's Rays." It was presented at the 1856 meeting of the American Association for the Advancement of Science (AAAS)—but not by her. Elisha read his own research paper, whereas Eunice's was introduced and read by

Henry. Unlike the other reports at that meeting, it was not published in the proceedings. Nevertheless, it became somewhat known. It was briefly described in a few newspapers. *Scientific American* wrote that Eunice's work offered "abundant evidence of the ability of woman to investigate any subject with originality and precision." Fortunately, it did appear in full under Foote's name in the 1856 issue of *The American Journal of Science and Arts*.

But then her discovery became lost. It might have remained so except for retired petroleum geologist Raymond Sorenson and a dose of what he calls "blind luck." He collected old copies of the *Annual of Scientific Discovery*, a yearly summary of scientific research. In the volume for 1857, Sorenson came across a description of Foote's work, and quickly realized that he was seeing the first reported connection between CO_2 and climate change. In 2011, he presented his findings in a journal published by his professional society. "I've had more response to that than anything else I've ever written," he has said, attention that continues as scientists and historians absorb the revelation of Foote's work—especially in 2019, the 200th year after her birth.

Foote's research used simple apparatus, two cylinders (presumably made of clear glass) 4 inches across by 30 inches long that each held two thermometers. She filled these tubes with different gases and compared their thermal responses to sunlight as judged by their temperature rises. Foote found that moist air absorbs more solar radiation than dry air; and what she called carbonic acid gas, that is CO_2, absorbs more radiation than ordinary air. That tube reached a temperature of 120 degrees Fahrenheit compared to 100 degrees Fahrenheit for ordinary air and cooled more slowly. She also found that the tube with CO_2 became much hotter than tubes filled with hydrogen or oxygen under equal conditions.

Foote unquestionably preceded Tyndall in discovering the major roles of CO_2 and water vapor. This was remarkable for an amateur from the young American scientific community, compared to an educated professional working within a mature European scientific setting. Tyndall, however, established a critical feature that Foote had not. Using an infrared source in his well-equipped lab, he showed the greenhouse effect is not triggered by direct sunlight but by infrared radiation arising from the Earth's warmed surface. Foote's experiment was not designed to distinguish between the two modes (although present-day analysis shows evidence of the infrared effect in her data). Tyndall deserves credit for uncovering this essential part of the greenhouse mechanism.

One lesson from Foote's life is that original research arises from access to a scientific education and opportunities for mentorship and discussions. Multiple studies show the same holds true today. It's also important to reflect on how Foote's research was received. Why couldn't she present her own work at AAAS, and why was it omitted from the published proceedings? Why didn't her work receive wider attention? As Leila McNeill underscored in Smithsonian.com, bias against women is a large part of the answer. Women could join the AAAS in that era, but science historian Margaret Rossiter notes an internal hierarchy in the organization that rated "professional" men over "amateur" women. This may account for the downgrading of Foote's presentation.

Later, Tyndall made a more serious omission. The papers he wrote about his discovery in 1859 and then in 1861—now five years after Foote's work—did not mention her results, although he cited those of Fourier and others. Priority of discovery has always been a serious matter in science. With no citation to Foote's work, Tyndall took center stage as "the father of climate science." Equally important in the long run, his omission also kept Foote's work from contributing to climate science as it developed in Europe.

Sexism colored Tyndall's perception of women as scientists, as revealed in a passage from the biography, *The Ascent of John Tyndall*. The book's author Roland Jackson writes that Tyndall "often exhibited surprise at women's intellectual capabilities, and though he imagined that women could understand anything revealed by the savants, he did not believe they had the same powers of imagination and discovery."

Was this attitude enough to make Tyndall purposely ignore Foote's work? In an article published this year, Jackson denies the possibility. He argues that poor communication between American and European science makes it likely that Tyndall simply did not know of Foote's research. He adds that covering up her work would not have been in Tyndall's character. Perlin, however, believes the omission was deliberate. He argues that Foote's research appeared in publications where Tyndall was virtually certain to have seen it. Perlin also notes a separate incident where Tyndall failed to credit earlier work by another American researcher (who happened to be Joseph Henry).

Whatever the link is between Tyndall and his treatment of Foote's research, his attitude echoed what many of his male colleagues thought about women in science. Despite this barrier, Foote is an inspiration for female scientists today. One of them, Leila Carvalho, is a professor of meteorology and climate sciences at the University of California, Santa Barbara. Responding to a recent seminar about Foote held at the university, Carvalho wrote that Foote's accomplishments "trembled my very core," and added, "I have been wondering how many Eunice Footes are out there to be discovered, and how much of their legacy has been veiled or even discredited because they cannot resist the pressure against gender, ethnicity, and race."

Carvalho underlines a lost opportunity for science. Had Foote's results been quickly and widely recognized in the U. S. and Europe, they may have generated American interest in climate science. That science could have been applied to climate change, as old tree rings and other data showed a rapid climb in Northern Hemisphere temperatures after the 1860s. With such a head start, the U. S. might have developed a deeper appreciation and a stronger response to the dangers of climate change than we see today from our government. This would benefit the whole world, and all of it coming from the efforts of an American woman scientist.

Romancing the Quantum

Dollars for science are a big topic these days. We hear about enormous budgets for the Superconducting Super Collider (now at $8 billion and rising)* and the space station, or for research agencies like the National Institutes of Health. These total $21 billion a year. It's natural for legislators and citizenry to wonder, why do those scientists ask for so much? As a participant in Washington's science scene, I know this is an important question. As a working research physicist, I know part of the answer: We ask because we are driven by the same romantic imperative that propels artists.

This insight came when I taught physics at a college of art. I worked hard to pull the art students into science, teaching them about light and optics, color theory and camera lenses, and showing them the pure colors of laser light. A deeper link yet, and my insight, came from a moment of self-revelation. One day I spoke of my life as a graduate student at the University of Pennsylvania in the 1960s, surviving on $3600 a year. I saw nods of recognition from the artists-to-be when I described how the poverty wasn't just unimportant; it seemed the essential romantic prelude to the field where my heart lay. Along with them, I grasped how my student life resembled the powerful image of the artist starving in a garret for art's sake, like a character in Puccini's *La Boheme*. (I even lived in a garret, but that's another story.) The image recurred later in my career, when I took a big pay cut to move to a research job that allowed greater self-expression.

This experience, among others, crystallized my suspicion that a romantic component exists in us scientists as surely as in poets and artists. Like them, we have deep inner urges, which prompt us to use mind and soul to express ourselves; we emulate heroic figures in our field, who define its highest aspirations; and we worry about the purity of our work, its congruence to our inner vision.

This does not mean that scientist and artist think alike. I'm speaking only of our common image of a quest, in which we play the romantic heroes. Ursula Le Guin writes of the "noble and austere" idea that "the artist must sacrifice himself to his work ...His responsibility is to his work alone. It is a motivating idea of the

Romantics...This heroic stance, the Gauguin Pose, has been taken....
as natural to the artist."

The inwardly focused Gauguin Pose is natural to the scientist
too. Here's Nobel Laureate Carlo Rubbia: "Detectors [devices
used to search for elementary particles] are really the way you
express yourself...[For] sculptors, it's sculpture. [For] experimental
physicists, it's detectors. The detector is the image of the guy
who designed it." Here's a generic quote from any of innumerable
scientists: "We're searching for the Holy Grail"—a room temperature
superconductor, the details of the human genome, or simply "the
truth." Not that personal needs don't exist alongside the romantic
ones. Noted scientist Victor Weisskopf comments that "a large dose
of ambition is mixed into the fervor – acclaim, tenure, a Nobel prize,"
but he too understands the underlying crusade.

If artists and scientists share the quest, we also share a need
for patronage—in 1991 terms, federal dollars from the likes of
the National Endowment for the Arts and the National Science
Foundation. But here science and art part company. The tools for the
artistic quest are cheap; the equipment to seek the Holy Grails of
science is too dear to outfit every possible search party.

So how should money be allocated among the scientific
Galahads? If you ask us, you encounter the quest mentality head on.
Each scientist's essential focus creates the belief that his or her work
is the most important in the world. As a result, scientists find it hard
to accept that society can provide only limited support. This reaction
has been criticized as a self-serving feeling of entitlement among
scientists. But it comes from commitment to a nobler ideal, that the
work must go on at any cost.

If the scientists' own criteria may be romantically colored, how
can we—all of us, not just the scientists—decide how many science
dollars are needed, and where they should go? First, we need to
accept a simple reality: Science is not hyper-human. It does not exist
in some transcendent space arching above human concerns. It is an
activity supported by humanity's resources, and properly subject to
broad human evaluation.

In this light, we should be uneasy about rankings of "cultural
value" that give high grades only to pure science, remote from any
application. It is an old tradition that the more basic the science, the

higher its standing. But boundaries between science and technology, between biological research and clinical practice, are now so blurred that these may no longer be useful distinctions. The creative tension between theory and tangible result, resembling the artistic interplay between beauty and function, can give deeply satisfying science. Semiconductor quantum wells—tiny structures made for nothing grander than electronic hardware—display beautiful manifestations of quantum mechanics, a highly abstract fundamental physical theory. Today's successes in fundamental elementary particle physics trace back partly to the Manhattan Project to build the atomic bomb, a great technological effort.

I have colleagues who choose not to deal with these issues. Some simply don't care if their work has impact in humanity's real world—after all, one never knows how and when breakthrough ideas come. The romantic quest is central to the meaning of science. It also keeps us scientists going in a profession which, with all its deep fulfillment and high prestige, is a demanding one. But the quest should not alienate scientists from society. Our expeditions into the unknown depart from society's shores; no matter how far out our journeys range, uncovering strange beasts and wondrous truths, we should return to those shores.

---

*The enormous Superconducting Super Collider was never completed. The project was cancelled by the U. S. Congress in 1993.

Technology in Society

Introduction

Technology was a main category in my first collection, *Real Scientists Don't Wear Ties*. It remains central and, in some ways, has become a bigger part of our lives since then. But now my treatment has a different slant due to COVID-19, and also due to the realization that supposedly value-free technology can embody principles that may be ethical but may also reflect immoral or biased viewpoints. It has become clear that technology cannot be considered in isolation, but only as it interacts with individuals and society. The topics in this section are about noteworthy technology and its roles in current issues.

Technology: past, present, and future

"Most Tech Today Would be Frivolous to Ancient Scientists" (2019) compares today's consumer tech to ancient technology, as seen by a classicist who brings views you will never see in Silicon Valley. "Sociophysics and Econophysics, the Future of Social Science?" (2018) discusses new mathematical tools that predict human behavior in crowds, voting blocs, and the marketplace. This kind of

Science Sketches: The Universe from Different Angles
Sidney Perkowitz
Copyright © 2022 Jenny Stanford Publishing Pte. Ltd.
ISBN 978-981-4877-94-7 (Hardcover), 978-1-003-27496-4 (eBook)
www.jennystanford.com

analysis is related to a worrisome trend toward judging people solely through algorithms. "When Houses Grow on Trees" (2009) answers a question about future materials technology sent to *Discover* magazine and ends up speculating about genetic engineering in plants. Today the biologically produced PHA plastics named in the piece are under intensive study, but I have yet to see anyone producing tree-grown plastic houses.

Robots, AI, and synthetic life

As robot technology reaches new heights, "Representing Robots: Theater First, Film Later" (2012) reminds us that the roots behind the very name "robot" allude to "worker" or "forced labor," as presented in the 1920 play *R. U. R.* An even older tale, about Dr. Frankenstein and his creature, must enter into any consideration of the ethical questions arising from the modification of people and the creation of life-like or living beings. "Frankenstein and Synthetic Life; Fiction, Science and Ethics" discusses Mary Shelley's story and other fictional accounts of artificial creatures, as well as scientific progress toward actually creating them and the resulting ethical issues. "Intelligent Machines: It's More Than Just Intelligence" (2018) raises similar issues for robots and AI. The piece speculates about why we find these creations so fascinating and shows that science fiction can help us understand our relationship to them and to ourselves.

Is technology neutral?

These pieces present ways in which nominally neutral technology can be put to corrupt uses or is explicitly biased. "Confronting the Wall" (2018) reviews films that use satellite technology to show what a massive undertaking it would be to build a wall between the U. S. and Mexico, but nevertheless also show how neurotechnology could make complete separation economically viable if morally questionable. "Bad Blood, Worse Ethics" (2018) reviews the book *Bad Blood*, about the downfall of the multi-billion dollar Silicon valley Theranos company, which deceived patients and investors about its high-tech blood tests. "The Bias in the Machine" (2020) uses the false arrest of an innocent Black man, and scientific evidence, to show that facial recognition algorithms used today by police departments are inherently biased against people of color. The piece suggests how this can be made right.

Technology and warfare

Technology has always played a role in warfare. "The War Science Waged" (1991), written during the First Gulf War against Iraq, relates how advances in semiconductor science, including my own type of physics research, made possible pin-point targeting in U. S. missile attacks, with the laudable goal of minimizing civilian casualties. That goal is said to also guide today's human-guided drone missile attacks in war zones. But as "From Terminator to Black Mirror: Algorithmic Warfare's Perils," (2020) points out, nevertheless drone attacks killed or injured over 200 civilians in 2019. Now, as the piece relates, there is the possibility of using facial recognition algorithms to enable smart weapons to select their own targets, with potentially far worse outcomes.

In an earlier time, America's nuclear weapons technology ended World War II. The morality of the atomic bombing then of Japanese civilians in Hiroshima and Nagasaki has long been discussed. "Radiations: What Came Out of the Atomic Bomb" (2020) reviews the book *Atomic Doctors*. Its contents add a new page to the discussion by revealing how the U. S. military and government tried to downplay the role of nuclear radiation in harming civilians, and how the story finally emerged.

Police technology

Among the methods that the police and the justice system use to find and convict criminals, forensic techniques have played a major role. Two recent reports, however, have pointed out problems in forensic practices. "The Physics of Blood Spatter" (2019) shows how flawed blood-spatter analysis has led to at least one unjustified imprisonment, and how the analysis can be improved through the principles of fluid mechanics. "From the Lab to the Courtroom" (2020) highlights the work of two physicists whose early efforts in ballistic analysis and other forensic areas helped solve two famous crimes in the 1920s and 1930s.

Most Tech Today Would be Frivolous to Ancient Scientists

Surrounded by advanced achievements in medicine, space exploration, and robotics, people can be forgiven for thinking our time boasts the best technology. So I was startled last year to hear Sarah Stroup, a professor of classics at the University of Washington, Seattle, give a speech called "Robots, Space Exploration, Death Rays, Brain Surgery, and Nanotechnology: STEMM in the Ancient World." Stroup has created a college course integrating classics and science to show how 2000-year-old Greek and Roman STEMM (science, technology, engineering, mathematics, medicine) underlie and illuminate the sciences today.

Stroup starts with robotics. The Greeks made self-acting machinery such as an automaton theater, a first step toward building a real robot, and they imagined a mythological one. Talos, a bronze being made by the god Hephaestus (later the Roman Vulcan) patrolled the island of Crete and threw rocks at threatening ships, anticipating today's development of intelligent battlefield weaponry that chooses its own targets. In the 4th century BCE, Aristotle foresaw other implications of intelligent machines when he wrote, "If every instrument could accomplish its own work... chief workmen would not want servants, nor masters slaves," as is now happening when robots and artificial intelligence replace people.

Stroup also gives the example of a "death ray," what the laser was called after its invention in 1960. In 214 BCE, the Greek scientist Archimedes may have created an early version. He is said to have used metal mirrors to focus sunlight that burned the wooden ships of the Roman fleet attacking his city, Syracuse. Though the story is probably apocryphal, modern researchers have shown that metal mirrors could concentrate sunlight sufficiently to set a wooden craft afire.

Presenting other accomplishments, Stroup notes that the Greeks explored space by naked eye and correctly concluded that the planets orbit the Sun. The anatomist Herophilus dissected the human brain to understand its structure, and Hippocrates established medical ethics, expressed today in the Hippocratic Oath that defines a physician's responsibilities. Centuries before modern nanotechnology, Roman artisans colored glass with embedded gold and silver particles only nanometers across, producing the famous Lycurgus Cup that looks green or red depending on how it is illuminated.

Stroup stresses that ancient STEMM both supported warfare and supported the common good. This dual nature exists today, and I was curious to learn how Stroup compared the uses and consequences of technology in ancient times to ours. Not long after we met at her talk, we conducted this interview over email.

We use the word "technology" daily. What would the word or concept have meant to an ancient Greek or Roman?

"Technology" comes from the Greek τέχνη, techne, which designates art, skill, or cunning. In Greek, it can be applied to sculpture, to metallurgy, to any craft or a method or set of rules for doing anything. The Latin translation of τέχνη would be *ars*, from which we get our word art. I find it amusing that moderns tend to imagine technology and art as opposites, when in fact the root words—*techne* and *ars*—mean exactly the same thing. In terms of τεχνολογία—*technologia*—it means specifically a systematic treatment of grammar. The modern sense of the word technology is not found in the ancient word.

In your presentation, you mentioned ancient technology versus what you call modern tech. How do they differ?

Much modern technology is built off ancient technologies, and in many cases, we still don't understand how they did certain things they did: We've not yet regained their knowledge. A major difference between ancient technology and modern tech is that the latter is industry driven, whereas ancient technologies never were. As a result, modern tech is designed not necessarily for use value—much modern tech is entirely frivolous—but for a consumer market, and is designed for early obsolescence. Luxury tech must become unusable as swiftly as the consumer will tolerate. Can you imagine an aqueduct that needed to be replaced every 18 months?

We can and do build durable technologies—medicine; aerospace—but the tech that most people depend on must appeal to our fears and vanities and must require continuous and rapid overturn. If it were truly *necessary*, the market would demand durability. Much modern tech is little more than current fashion, of which moderns have become the passive consumers.

What do students take away from your course?

The students generally experience a kind of existential crisis at some point in my class once they realize how much was lost from antiquity and how relatively little the modern period has produced. We've rediscovered a good deal, and of course the control of electricity has helped tremendously, but they had developed atomic theory, and I don't think they were too far away from being able to control electricity.

We talk about how moderns have trouble dealing with the advanced technologies of the ancients—we say things like ahead of his time, but if all these guys were ahead of their time, what you really mean is that was their time. As one student said last week, it's like the typical "undeveloped society sees advanced technology of more sophisticated society, suspects magic," except in this case, we're the ones suspecting magic.

Besides the example of Hippocrates, how do your students learn about ethics as it connects to technology?

Ethics is a Greek invention. Aristotle was the first philosopher we know who began to write treatises concerned with what came to be known as ethical philosophy. Ethics—and philosophy as a whole— were born along with Greek astronomy, mathematics, science, and engineering.

Your students build ancient devices, like catapults, as part of your course. What does that teach them?

They realize that so much ancient engineering, like modern engineering, is about building things to kill people. We all have great fun launching catapults, trebuchets, and ballistas, but we also talk about the fact that each is a weapon and was designed to kill. The main thing the students learn is how difficult it is to build these things. You spend nine weeks trying to build a catapult—with access to a huge workshop and electric tools, and you can use nylon cord— and it still barely launches a ball more than 10 feet. Suddenly the past doesn't seem so laughably simplistic. Though by the end of the term, they're all over that misconception, anyway.

How do you understand the value of history?

History is a roadmap, and if you don't study it, you have only a scrap of paper with "you are here" scrawled on it. The more you learn about history, the larger your map. The larger your map, the more you are able to know where you've come from, and so know where you're going.

As a classicist, what inspired you to develop a course in ancient STEMM?

My degrees are in philosophy and classics, but my background is in the sciences and that's what I had planned to study in university. I read in the sciences, I design and build high power rockets, and I adore math. As it happened, I stumbled into philosophy and Greek, the first two things to challenge me, and so I ended up staying here.

While we know a fair deal about ancient Greek and Roman technological sophistication, there weren't any courses in ancient technology/science/math when I was an undergrad. "Ancient technology" had never been part of the modern canon "classical studies," and the state of the evidence is even worse than for the rest of our field. Literary masterpieces make it through time, but technological treatises are disposable, and advanced machinery is melted down and reused. I invented my course both to fill this void and to give myself the excuse to study more of this myself. Also, I became weary of modern STEM sorts, imagining they had invented everything.

In your talk, you noted the Greeks used a recyclable material, bronze. What would they have thought about our global dependence on plastic, a material used once, then discarded, but never again to go away?

I believe they would be absolutely appalled.

Sociophysics and Econophysics, the Future of Social Science?

In 2014, Stephen Guy of the University of Minnesota and his colleagues described how people move to avoid hitting each other when interacting in large groups. "Human crowds," they wrote, "bear a striking resemblance to interacting particle systems." Pedestrians move, the researchers observed, like negatively charged electrons, which repel each other more strongly as they approach, with one key difference. Unlike electrons, pedestrians anticipate when a collision is imminent and change their motion beforehand by swinging wide to avoid a crash.

Using this knowledge, the researchers derived a mathematical rule for an electron-like "repulsive force" between any two pedestrians but based on time-to-collision rather than distance. This allowed the researchers to correctly predict how a moving crowd bunches up when funneled into a narrow passage, or spontaneously forms directional lanes, as when football fans leave a stadium headed for different exits. Other *sociophysicists* have applied similar principles to auto traffic.

The idea that math and physics can illuminate human conduct dates back to the 18th-century Scottish philosopher David Hume. Later, the French philosopher Auguste Comte proposed that there are general laws describing human societies, and the Belgian mathematician Adolphe Quetelet began performing statistical analyses of human qualities. Today the sciences of *sociophysics* and *econophysics* draw on all these ideas in their attempts to explain human behavior.

Some of this work uses the methods of statistical physics, which studies how swarms of particles interact to produce new effects. For example, individual H_2O molecules move randomly in water. But, cooled to zero Celsius, they undergo a "phase transition" and lock together into solid ice. Similarly, whatever the varied qualities of individual people, they can form voting blocs, which may spontaneously reform after a crucial event, such as a debate between political candidates. The French scientist Serge Galam applies physical effects like these to human behavior in his book *Sociophysics* (2012).

Another approach to sociophysics extracts mathematical rules and behavioral patterns from terabytes of data gleaned from existing records or from digital interactions like social media. *Social Physics* (2015), by MIT researcher Alex Pentland, presents this data-based method as a source of "reliable, mathematical connections between information and idea flow on the one hand and people's behavior on the other."

In truth, meaningful sociophysics necessitates both data and a model or theory. As sociophysicist Frank Schweitzer of ETH Zürich points out, analyzing data may uncover interesting and useful correlations, but it will not produce "an understanding of causal relationships... Successful sociophysics models tend to have interfaces with both empirical data and social theories."

An early example of this approach in practice came from Princeton astrophysicist turned sociophysicist John Q. Stewart. He noted in 1948 that for many kinds of interactions between groups of people, such as telephone calls between any two cities in the U. S., the number of interactions is proportional to the product of the group populations divided by the physical distance between them— that is, more interactions for bigger groups and smaller distances. This seems intuitive, especially since phone companies charged callers extra for "long distance" dialing; but Stewart also saw that the mathematical equation is like that for Isaac Newton's expression for the gravitational energy between two objects, which is proportional to the product of their masses divided by the distance between them. From this analogy, Stewart defined a "demographic energy," an invisible force that structured the distribution of U. S. populations across urban and rural areas.

In Stewart's time, most scientists found such analyses to be of little value. These interdisciplinary approaches are still less widely accepted than traditional physics, sociology, and economics. But with new analytical tools and new data, researchers are finding convincing results.

In 2011, David Garcia and Frank Schweitzer of ETH Zürich studied how people influence each other by examining nearly 2 million anonymous Amazon reviews of thousands of books and products. They did this within the "circumflex" model of emotions, where an emotion is defined by: (1) valence, the degree of pleasure or pain associated with it; and (2) arousal, the activity prompted by the emotion. These qualities are believed to arise from two independent human neurophysiological systems.

The researchers assigned a "sentiment" score from −5 to +5, highly negative to highly positive, to the emotional content of each Amazon review. They also examined the actions induced by the sentiments, such as when people rate reviews as "helpful" or "unhelpful" or are moved to write their own reviews, which also carry emotion. The results clearly show that individual reviewers are influenced by other comments, and that unlike negative reviews, positive reviews skew toward highly positive—apparently people who like something want to show that they like it a lot. By comparing reviews of one of the Harry Potter books, which received much media attention, to reviews of another book that did not, the researchers found distinct differences in group emotional patterns, reflecting different emotional responses to external marketing versus internal word of mouth.

Results like these have obvious value for corporate marketers. But on a larger scale, studying how opinions travel through groups is an urgent matter for politics, leading sociophysicists to study "opinion dynamics" with the methods of statistical physics.

In 2015, a team at the City College of New York (and other institutions in the U. S., Brazil, and Israel) published its research on extreme opinions, noting the "worldwide trend towards the division of public opinions about... political views, immigration, biotechnology applications, global warming" and more, including cultural matters. "A marked dwindling of moderate voices is found," they write, "with the concomitant rising of extreme opinions... the opinion or attitude of an initially small group could become the rule." The authors analyzed hundreds of survey results from many countries, where people gave their opinions about religion, politics, abortion, and other subjects as "very favorable" or "very unfavorable," classified as extreme; or as "somewhat favorable" or "somewhat unfavorable," classified as moderate.

Comparing the fraction of respondents f_e with extreme views to the fraction f who held any opinion at all, the researchers found a surprising pattern. When there were just a few extremists, f_e was proportional to the total number of opinion holders, as expected for a group of non-interacting members. But at a certain critical value of f_e, around 20% of the respondents, f_e plotted against f began rising in a steep nonlinear manner that would lead quickly to a majority of extremists. When there are enough extremists to interact with each other, it amplifies their impact.

This effect is central in statistical physics, where correlations among particles produce nonlinear behavior that cascades into a phase transition, like water turning into ice. When the researchers modeled the group interactions, including an element of "stubbornness," resistance to changing an opinion, they were able to reproduce the empirical data and also to develop a social "phase diagram." Analogous to a diagram that shows the conditions of temperature and pressure under which water exists in a solid, liquid, or vapor phase, this diagram shows the conditions for a society to occupy moderate, incipient extremist, or extremist phases.

Quantitative methods have been applied to other social phenomena, such as the spread of rumors and voting patterns in recent European and U. S. elections (though no research I know of predicted Donald Trump's presidential win in 2016) and are useful in economics as well. According to the authors of *Econophysics: An Emerging Discipline*, econophysics received a strong push after traditional economics failed to foresee the global economic crisis of 2008. That lack, they claim, shows that rather than relying on theories that may be too simplified to describe reality, economics needs an injection of econophysics, with greater weight given to empirical data.

One current interest of econophysicists is mathematical power laws, where one variable depends on another raised to some power. For example, the area of a square and the volume of a cube are given by the length of a side L raised to the power 2 (L^2) and power 3 (L^3), respectively. For reasons that are not yet completely understood, power laws describe a range of economic activities, such as stock market transactions and, especially relevant today, income inequality. Since the late 19th century, data analysis has shown that the fraction of people with an income greater than some cutoff value follows a power law called the Pareto distribution, which gives a small slice of the population a disproportionately large share of income and wealth. In 2016, for instance, the richest 1% of U. S. households owned 40% of the nation's wealth, a very high level of inequality.

The question of global inequality was brought into focus through the groundbreaking 2014 book *Capital in the Twenty-First Century* by the French economist Thomas Piketty, whose analysis includes the Pareto distribution. In 2015, Stanford University economist Charles Jones wrote in his article "Pareto and Piketty" that this distribution is a "key link between data and theory." If econophysics can provide deeper understanding of the origins and meaning of the Pareto distribution, it would greatly contribute to understanding the important issue of inequality.

Sociophysics and econophysics are answering questions about how people behave, though no one could yet claim that these approaches are uncovering deep truths about human nature. In a 2012 essay for the *Guardian*, the Nobel Laureate economist and *New York Times* columnist Paul Krugman wrote about the experience of reading Isaac Asimov's *Foundation* trilogy as a teenager. The Foundation books concern a future scientist, Hari Seldon, who invents the science of "psychohistory." The equations of psychohistory, which describe how human societies evolve, predict the coming fall of the Galactic Empire. Armed with this knowledge, Seldon creates the Foundation, a group devoted to minimizing the dark times, so society can recover and go on. *Foundation* greatly affected Krugman, apparently. "I grew up wanting to be Hari Seldon, using my understanding of the mathematics of human behavior to save civilization," he wrote.

We are still a long way from the elegance and power of famous results in physics like Isaac Newton's equation $F = ma$ or Einstein's $E = mc^2$. But it took millennia for physicists to derive these insights. Maybe in only a few more centuries, we will become like Hari Seldon, able to better understand ourselves through quantitative science.

When Houses Grow on Trees

Editor's note from *Discover* Magazine's *Codex Futurius* blog: Yes. It's true. After a little summer slow-down, it is time for the return of the *Codex Futurius*, this blog's never-ending quest to explore the big science of science fiction. This question on futuristic materials was fielded by Sidney Perkowitz, a physicist at Emory University. Thanks much to Dr. Perkowitz for the solid (ha) info and to Jennifer Ouellette, the director the NAS' Science and Entertainment Exchange program, for connecting us with him.

Will we use metal in the future? What else would we build things out of? Might we use organic technology (machines and buildings made of or from biological organisms) instead?

In *The Graduate*, that iconic film from 1967, bewildered 20-something Benjamin Braddock (Dustin Hoffman) gets some career advice from a businessman who leans close and intones "I want to say one word to you. Just one word. Are you listening? Plastics." Benjamin didn't follow that advice, but the rest of the world did, and in spades. By 1979, global production of plastic had exceeded that of steel and is still growing, reaching over 200 million tons this year.* There's no doubt that plastic will continue to play a major role in how we make things, but it won't replace everything.

In some ways, plastic is the material of the future, the latest step in humanity's long upward trek through the ages of stone, bronze, iron, and steel. The word "plastic" comes from Greek roots meaning "capable of being molded." Compared to metals and other materials, plastic is infinitely versatile. With its ability to shape-shift and to take on different mechanical and optical properties, it shows up in a huge spectrum of applications from packaging and plumbing to toys, medical supplies, and computers. And unlike iron and steel, plastic doesn't rust. But plastic also has problems that will prevent it from replacing metals any time soon. Its very durability can be an issue. Discarded plastic objects can survive for centuries in garbage landfills without degrading, and plastic artifacts have been found polluting the oceans far distant from any land. Also, what doesn't seem to be widely appreciated, the raw material to make plastic comes from a resource we need to conserve, petroleum.

On top of this, metals do some things better than plastic—just try cutting up an apple with a plastic knife. Copper and other metals are needed to conduct electricity through power grids; all plastic can do is insulate the current-carrying wires. However, plastic is making inroads relative to some materials such as wood, which is being replaced by plastic "lumber" in certain applications. Plastic also offers a possible way to actually construct things using biotechnology. Unlike metals, which are classified as inorganic, plastics are organic; they're made of carbon, hydrogen, nitrogen, and oxygen, the same constituents as living things, which links plastic to biological products.

For instance, under the right conditions, certain microorganisms can synthesize compounds called polyhydroxyalkanoates (PHAs). These display properties like those of artificial plastics, with the benefits that they're not petroleum-based and are biodegradable. Researchers are investigating ways to mass produce these bioplastics, for instance by bioengineering plants to create them.

If you want to speculate even further, way past the idea of growing plastic rather than making it in factories, think about the possibility of bioengineering plants as in science fiction to produce plastic exactly in a desired shape, from a drinking cup to a house. Current biotechnology is far short of this possibility, but science fiction has a way of pointing to the future. If bioplastics are the materials breakthrough of the 21st century, houses grown from seeds may be the breakthrough of the 22nd.**

---

*That was in 2009. By 2019 the production of plastic had doubled to over 400 million tons.

As another way to reduce the use of plastics, MIT scientists announced in 2021 that it might be possible to grow wood-like plant tissue into useful shapes such as boards or more intricate structures: A. Beckwith, J. Borenstein, and L. Velásquez-García, Tunable plant-based materials via in vitro cell culture using a *Zinnia elegans* model, *J. Cleaner Production* **288, 15 March 2021, 125571.

Representing Robots: Theater First, Film Later

Reviews of the plays *R. U. R.* (1921) and *Gizmo* (2012), and the film *Metropolis* (1927) and others.

When I made a list of the all-time 10 best science fiction films for my book *Hollywood Science* (2010), I was surprised to find that three of them feature artificial creatures: machine-like robots in *Metropolis* (1927) and *The Day the Earth Stood Still* (1951), and human-like androids in *Blade Runner* (1982). Artificial beings are big in other science fiction films too. A keyword search on "robot" in the Internet Movie Database yields hundreds of feature films, from *The Master Mystery* (1920) through *Westworld* (1973), *RoboCop* (1987), and *A. I.* (2001) right up to *Real Steel* (2011) and this year's *Prometheus*, with more in production.

But despite the many robot films, stage rather than screen was seminal in establishing robots themselves. In 1921, the groundbreaking play *R. U. R.* (*Rossum's Universal Robots*) by Czech playwright Karel Čapek introduced the idea of artificial beings manufactured to serve humanity. These were dubbed "Robots," from "robotnik," Czech for "worker." The play became a worldwide sensation after it opened in Prague and has been frequently revived, as recently as 2011. In describing what happens as the robots become more nearly human, *R. U. R.* raises questions that were relevant in 1921 and are even more so now, when technology is bringing us that much closer to creating artificial beings.

Film caught up six years later when *Metropolis* posed similar questions. Its robot is a striking, unsettling merger of metal body and womanly features, built by the scientist Rotwang to replace the woman he loved and lost. The boundary between nonhuman and human blurs further when Rotwang turns the mechanical robot into a physical copy of a real woman named Maria—but not a psychological or spiritual copy, for the android is lewd and treacherous, not chaste and noble like the real Maria. Rotwang goes on to suggest that people will be completely replaced by robots, "the workers of the future...now we have no further use for living workers." That theme also shows up dramatically in *R. U. R.* when the robots wipe out humanity in a violent uprising.

Between them, *R. U. R.* and *Metropolis* grapple with the differences between real and manufactured people, and with the morality of creating and exploiting lifeforms. This is in the tradition of thoughtful science fiction that foresees scientific developments and where they will take us. But later, theater and film took different paths in depicting robots. Hollywood's science fiction films have become major money makers that provide popular platforms for stories about robots; but though theater has recently embraced plays about real science such as Michael Frayn's Copenhagen, it has done less with science fiction or robots.

One important difference is that movies can convincingly display robots in ways that theater cannot. In early films, robots like Gort in *The Day the Earth Stood Still* (*TDTESS*) were more or less well played by actors in costume. But when digital special effects were first used, in *Westworld*, it was to show how the world would look as seen by a robot. Now computer-generated imagery (CGI) routinely creates dynamic robots on screen. In the 2008 remake of *TDTESS*, CGI produced a spectacular Gort that could separate into a swarm of nanomachines. In *Real Steel*, set in a near future where robot boxing is a major sport, CGI created action-filled scenes of robots smashing each other into junk.

But good stories do not live by CGI alone. They need a compelling plot and characters. If they also convey important ideas or a message, so much the better. The original 1951 *TDTESS* had a tight plot with considerable action, appealing characters, and a meaningful message for that Cold War era—namely, be very careful with nuclear weapons. This earlier version is far less spectacular than the later one, but it is a better film.

Theater, too, can provide rich experiences that do not depend on stunning visuals. Though some stage works are designed as spectacles—for example, *Spider-man: Turn Off the Dark* now on Broadway—stage dramas focus on people and their relationships, and in science plays like *Copenhagen*, on ideas. Karel Čapek understood this. Two years after the robots he invented appeared in *R. U. R.*, he wrote, "For myself, I confess that as the author I was much more interested in men than in Robots."

But since *R. U. R.*, the power of theater has hardly been used to broadly examine robot–human interactions, though a few plays have tackled some aspects. Replacing people with machines was treated in Elmer Rice's *The Adding Machine* (1923), and androids have roles in Alan Ayckbourn's *Henceforward* (1987) and Elizabeth Meriwether's *Heddatron* (2006), but none of these plays comment generally about robots within society.

However, Anthony Clarvoe's play *Gizmo*, first performed at Penn State's Centre Stage in April 2012, does just that. Inspired by *R. U. R.*, it is updated for our time and slightly beyond. As its script explains, gizmos are "flesh mechanisms, manufactured to serve...They do not aspire to be real live boys. They are neither cute, nor sentimental, nor evil. They are other than that." Comparing successively more human-like generations of gizmos and exploring the relations between gizmos and the people who create, befriend, or exploit them, *Gizmo* shows how living with "almost" humans would change us. As in *R. U. R.,* the gizmos triumph in the end—not violently but subtly, as humanity becomes dependent on them while adapting to a hi-tech world.

Like Čapek, Clarvoe explores people through their relation to technology, rather than examining technology itself. Dan Carter, who directed the Centre Stage production, puts it this way: "[*Gizmo*] is not *Transformers* on stage...we can't do that nor do we want to... ultimately this is a play about what it means to be human...we want to see ourselves on stage, we want to see people interacting with [the gizmos]." The human–robot connection is sharply drawn in theater, for to see a robot on stage is to watch a real person play an artificial being that in turn resembles a real person. Only live performance can bring that built-in experience of different levels of humanity and near-humanity.

A filmgoer watching a robot on screen, especially one generated by computer as used in CGI, has a different experience that lacks the living comparison to humanity. When CGI produces marvelous imaginary robots and even whole imaginary worlds, it is a huge asset for science fiction, but a film that relies too much on the power of CGI can lose the human story.

Still, robot films can present evocative insights. *Blade Runner* is a meditation on mortality. Its androids or "replicants" are full of rage and despair because they know they have been made with built-in termination dates. As I wrote in *Hollywood Science*, "Knowledge of death spiritually separates humans from animals. That same knowledge brings replicants...closer to humanity, and perhaps closer to possessing a soul..." In *RoboCop*, the brain of deceased police officer Alex Murphy is put into a steel body. Murphy's human instincts complement his enhanced physical abilities to make the cyborg an ideal cop, and Murphy eventually comes to terms with his man–machine status. The central question in *A. I.*, also posed in *Blade Runner*, goes back to *Metropolis*: could an android inspire love? And in *Prometheus*, the android David relates himself to us, his creators, as the people in the story relate to the alien or divine creators they seek.

A new *RoboCop* film came in 2013, and other robot films will keep coming. Even with CGI, some will tell affecting human stories or raise intriguing questions. Nevertheless, what theater can do remains unique. After *Gizmo*, perhaps playwrights will further treat artificial beings in the form of robots, or cloned people, or people modified by genetic, mechanical, or digital means. Theater does not reach millions as film does, but in its thoughtful and provocative way, it can help us understand what technology is doing for us and to us.

Frankenstein and Synthetic Life: Fiction, Science, and Ethics

Introduction

The creation of synthetic life may seem a project that could be realized in a 21st-century laboratory or may seem merely a fiction that began with Mary Shelley's story about Victor Frankenstein and his creature. In reality, making synthetic life is both science and fiction and even science and myth, with roots far older than this century or the 19th, when Shelley wrote. She understood this, for the full title of her story is *Frankenstein, or, the Modern Prometheus.* In Greek mythology, Prometheus was a Titan, one of the powerful divinities who preceded Zeus and the other Olympian gods. In some versions of the myth, Prometheus creates humankind, and in all versions he steals fire from the Olympians for the benefit of humanity. These actions angered Zeus, who punished Prometheus by sentencing him to eternal torment.

In Mary Shelley's novel, Victor Frankenstein creates a living being with the hope of creating a superior version of humanity, but like Prometheus, he pays a price for his aspirations. Stories like these raise timeless issues and express significant truths. At the deepest level, they reflect our own feelings about life and death. Furthermore, they present the long-standing moral questions that synthetic life would raise. A modern scientist could still share the same goals and fears as Victor Frankenstein, but with one great difference: now we truly have the scientific tools to modify and perhaps even create living things.

Some observers believe that life has already been synthesized in the form of bacteria that have been radically altered in the laboratory. Scientists have also already attempted to genetically modify human embryos. This effort was unsuccessful, but the scientific community took seriously the possibility of engineering genetic changes in humanity that could be inherited, and called for a moratorium on such work until we better understand its consequences.* These could be unpredictable for future generations and could lead to serious abuses by governments, supporters of the discarded idea of eugenics, special interest groups, or commercial enterprises. Yet if this same technology were to obtain wide approval and were wisely applied, it could give humanity better health, greater longevity, and increased abundance, and improve us as a species.

At the 200th anniversary of Shelley's story about the creation of synthetic life, we need to understand the science that could now make these different outcomes possible. Fiction and myth offer a first step. Long before there was a science of synthetic life, artificial lifeforms were being imagined in myth and legend, then later in literature and the theater, and finally in film, television, and the Internet.

These efforts, including Mary Shelley's story, add up to a fantasy history of synthetic life that illustrates the motivations for creating it, the means to achieve it and the deep ethical concerns it raises, providing an introduction to the real science of synthetic life and how society might respond to it.

Imaginary synthetic life

Since humanity's early days, different cultures have wondered about the beginnings of life and have expressed the desire to create it, typically in human form. Why? Perhaps to challenge the gods, or to outwit death, or to improve humanity, or simply driven by the so-called technological imperative: "because we can." When science had advanced sufficiently, its practitioners began trying to imitate, modify, or create life by different means—first mechanically in the form of early automata and robots and now within the emerging sciences of genetic engineering and synthetic biology.

At the same time, writers and thinkers imagined how humankind could create life. Often these imaginings expressed deep concerns about replacing God or the gods as the creators of life, which have different roots. One is based on fear: divinities reserve for themselves the power to make life, and like Zeus, become angry when humanity challenges this right. In the Judeo–Christian tradition, another factor is that God created humanity "in His own image," giving human life a unique value and sanctity. A different view sidesteps such issues. It is the purely materialistic and scientific approach that holds that during the multi-billion-year history of the universe and our planet, atoms became simple molecules and then complex ones that self-replicated. In a complicated process whose origins and details science does not yet fully understand, these eventually took on the characteristics we associate with "life" to become all the lifeforms we now know.

The history of imaginary synthetic life embraces both approaches. An early example of the divine creation of a living being is the Greek legend of Pygmalion, made famous in the work of the Roman poet Ovid. Pygmalion was a sculptor who made a statue of a beautiful woman. He fell in love with this ivory imitation, and one day, he found that it had turned into a real woman, brought to life by the Goddess Aphrodite in response to his wish.

Greek mythology also shows a god using technology to create what can only be called the first robotic synthetic being. The god was Hephaestus (Vulcan in Roman mythology) who worked with fire and metals to make devices such as Apollo's chariot. His greatest artifact was Talos, a giant bronze man-like construction that defended the island of Crete from enemy ships by hurling rocks at them. Talos' metal body, however, still needed a divine spark in the form of ichor, the blood of the gods, that ran through a vein in its body. When the ichor drained out of the vein, Talos was destroyed.

Other early metallic creations were also said to have life-like qualities. Two 13th-century religious figures and philosophers, Albertus Magnus and Roger Bacon, were each reputed to have made a talking head out of brass. Later a more humble material, clay, became the source for the golem, a synthetic creature from Jewish legend. The most famous of these was said to have been created in the 16th century by the rabbi of Prague to protect the city's Jews from pogroms, that is, violent organized attacks directed at them. This artificial being required another kind of divine spark. In the Old Testament, God fashions Adam from dust or clay and breathes life into him. Similarly, after the rabbi constructs the golem, it is animated by calling on the power of God.

One feature of the golem legend would continue throughout the saga of imaginary synthetic creatures. Though made to protect the Jews, the creature eventually goes violently out of control and must be deactivated. The idea that synthetic beings might run amok, turn against their creators, or be made for evil purposes raises fears that persist as we approach the possibility of really making them. Frankenstein's creature follows this tradition in turning on its maker.

In one crucial aspect, however, Mary Shelley's vision departs from earlier ones: Dr. Frankenstein makes his creature purely through scientific knowledge rather than by resorting to supernatural means.

Shelley's story gives few details about the creation of this synthetic creature—the Being, as Mary's husband Percy called it— but does tell us that it is made of parts taken from "the dissecting room and the slaughterhouse." And though the Being is ugly, it has some favorable features. Its limbs are in proportion, it is agile, and it displays a good brain as it speaks eloquently and reads classic literary works. (Director James Whale's definitive 1931 film version of *Frankenstein* makes Boris Karloff into a more monstrous Being. Its brain comes from a dead murderer, it emits only animal-like cries, and it lurches rather than walks).

Shelley's fictional Being with its intellectual and physical capacities, made not by God but by science, follows new discoveries of her time. In 1780, the Italian scientist Luigi Galvani had observed that the muscles in a frog's legs twitched when a voltage was applied. His discovery of what Galvani called "animal electricity" and is now known as bioelectricity generated huge interest before either electricity or life processes were well understood. It was not then a big step to believe that electricity might reanimate the dead, and experiments were made to revive hanged criminals by applying a voltage to the dead body (as described elsewhere in this anthology)—needless to say, unsuccessfully. Nevertheless, Shelley's introduction to the 1831 edition of *Frankenstein* alludes to these beliefs, stating, "Perhaps a corpse would be reanimated; galvanism had given token of such things."

Certainly, Victor Frankenstein has the background to perform a scientific reanimation. Shelley writes that as a boy Victor studied electrical science and moved on to chemistry and anatomy, using this knowledge to learn how to animate dead matter. After constructing the Being on a "dreary night in November," he collects "the instruments of life...that I might infuse a spark of being into the lifeless thing that lay at my feet." Shelley does not tell us what these instruments are, but clearly this is a scientific process.

Yet for all the neutral scientific approach, something deeper pulls at Victor when he beholds the Being he has created. After animating it, Victor says:

I had worked hard for nearly two years, for the sole purpose of infusing life into an inanimate body...I had deprived myself of rest and health. I had desired it with an ardour that far exceeded moderation; but now that I had finished, the beauty of the dream vanished, and breathless horror and disgust filled my heart. Unable to endure the aspect of the being I had created, I rushed out of the room...

Similar emotions appear in James Whale's 1931 film *Frankenstein.* When Dr. Frankenstein (Colin Clive) first sees the body that he has energized with a lightning bolt actually twitch, he is overcome, crying with a mixture of hysteria, exultation, and fear, "It's alive! *It's Alive*! IT'S ALIVE!...Now I know what it feels like to be God!"

These visceral reactions represent a realization that, despite the scientific trappings, the creator of synthetic life may sense that he or she is transgressing forbidden boundaries—simultaneously feeling pride in the achievement and fear that it will be punished. There are psychological factors too, as noted by the pioneering psychoanalyst Sigmund Freud in 1919. In his essay "The Uncanny," Freud wrote that humans feel dread in the presence of a dead body because that awakens hidden but intense feelings about our own mortality. Yet once we accept that a being has reached that final stage of nonexistence, it is a new shock to see it re-cross the boundary from death to life, becoming a person and not a thing—the shock that Dr. Frankenstein feels when he says "It's alive!"

Freud's "uncanny" border between living and nonliving is echoed in a modern possibility, the creation of robots that look and act human—that is, androids, which if they were self-conscious would represent a type of synthetic being. How human-like they seem depends on our reactions, because much of our willingness to see a synthetic being as truly alive depends on projecting our own characteristics onto it. In 1970, that led the Japanese robotics professor Masahiro Mori to the concept of the "Uncanny Valley." As robots evolve from lumbering metal machines to human-like appearance and behavior, people make emotional connections with them. But if an android becomes nearly but not quite indistinguishable from a person, empathy turns into revulsion, like a dip or "valley" in the empathetic link to the android. This may be why Dr. Frankenstein feels horror at his creation: It is somewhat human, yet not human enough.

Other fictional synthetic creatures appeared in the 19th century, starting just before *Frankenstein* was first published. In 1817, the German author E. T. A. Hoffman wrote "The Sandman," a Pygmalion story about a young man who falls in love with an imitation woman called Olympia, a clockwork automaton or early robot. She and similar artificial beings appear in the ballets *Coppélia* (1870) and Tchaikovsky's *Nutcracker* (1892), and the opera *The Tales of Hoffman* (1881).

Only in the early 20th century did really widely known synthetic beings appear, in the stage play *R. U. R.* and the film *Metropolis,* both of which introduced beings made for use by humankind. Karel Capek's *R. U. R. (Rossum's Universal Robots)* became a worldwide sensation after its premiere in Prague in 1921. It remains famous today because it gave us the word "robot," which derives from the word "robota" that means "forced labor" in Capek's native Czech and describes these entities made only to work for humanity.

The robots were made of an artificial organic material that "behaved exactly like living matter." It was used in "vats for the preparation of liver, brains...and a spinning mill for weaving nerves and veins" to mimic human bodies. Decades before the discovery of the role and structure of DNA, before genetic engineering and synthetic biology, Capek foresaw the creation of organic rather than mechanical synthetic humans and also gave them reason and emotion. They resent their slave status and desire revenge against humanity but also display the capacity for love.

Another enslaved being appeared in the remarkable film *Metropolis* (1927, Fritz Lang) that portrayed a futuristic city of that name. The story features a noticeably feminine mechanical robot built by the scientist-wizard Rotwang (played by Rudolf Klein-Rogge), as the first of many that will replace human workers like the robots in *R. U. R.* Later Rotwang transforms the robot into what we would now call an android. This is an exact physical replica of Maria, a real woman, but it lacks her moral sense (the robot, the real Maria, and her android copy are all played by Brigitte Helm). Through their synthetic beings that are more or less human, *R. U. R.* and *Metropolis* raise the still relevant moral questions, how should we treat our own creations, and its corollary, how would they treat us?

Other imaginary synthetic lifeforms followed. Many were mechanical robots, most famously in Isaac Asimov's science fiction story collection *I, Robot* (1950), which introduced the Three Laws of Robotics, a guide to robotic behavior toward humans (there was no corresponding guide for human behavior toward robots). Other media creations included the mechanical robots Gort and Robby in the films *The Day the Earth Stood Still* (1951, Robert Wise) and *Forbidden Planet* (1956, Fred M. Wilcox), respectively.

Mechanical entities like these could never be mistaken for people, but later two stories showed manufactured synthetic beings made to look and act human, that is, androids. In Ridley Scott's classic futuristic cult film *Blade Runner* (1982, based on Philip K. Dick's 1968 novel *Do Androids Dream of Electric Sheep?*), engineered "replicants" can pass as human but nevertheless know they are artificial with a deliberately limited lifetime. Led by the exceptionally strong and intelligent replicant Roy Batty (Rutger Hauer), they display the human trait of self-preservation. They violently rebel against their preset termination until a special policeman, Blade Runner Rick Deckard (Harrison Ford), destroys them. Another replicant, the advanced model Rachael (Sean Young), has a deeper level of self-awareness. She believes she is human and has difficulty in accepting that she is not.

Another significant android character is Lieutenant Commander Data (Brent Spiner) in the television series *Star Trek: The Next Generation* (1987–94). Like Roy Batty, Data is in many ways a physically and mentally superior being. As an officer in the United Federation's Star Fleet, he earns the respect and friendship of his human shipmates. Though initially he does not experience human emotions, an "emotion chip" implanted into his artificial brain later helps him to do so, but he still struggles with the complexities of human feelings and behavior.

The quandaries that Roy Batty, Rachael, and Commander Data face illustrate the difficulties in creating engineered synthetic beings. Even with a degree of self-awareness, their inability to become fully "human" highlight how hard it would be to artificially reproduce self-consciousness and the human condition, let alone create beings with superior abilities.

There is another approach, which is to modify or create life, including human life, through genetic manipulation. Once such efforts to "improve" humanity came under the heading of eugenics, the social philosophy that aims to promote reproduction among "desirable" people while preventing the "undesirable" from reproducing. This approach is now discredited, not least because of the vile role it played under the Nazis. Instead, we are considering how to genetically change people for laudable goals like fighting disease, and with the added realization that genetic manipulation is the modern science that might actually also be able to create new beings.

As fiction has explored synthetic humans, it has explored the consequences of genetic manipulation. In 1931, Aldous Huxley's novel *Brave New World* described a future society based on an advanced form of eugenics, where reproductive technology is used to put people into five castes from the highest Alphas to the lowest Epsilons. To create these categories, fetuses are cloned and chemically altered so they develop into people with either enhanced or limited abilities. The former are dominant; the latter are fit only for inferior roles in a stagnant, shallow, and corrupt society.

Decades later, writer-director Andrew Niccol's film *Gattaca* (1997), which NASA called the most plausible science fiction film ever made, foresaw a future dominated by eugenics and DNA (the title *Gattaca* combines the initials of guanine, adenine, thymine, and cytosine, the compounds crucial to how DNA functions). In a society focused on people with the "best" genes, the wealthy pay for enhanced "designer children" while genetically "inferior" people face limited lives and prospects. The story, however, has a redemptive element. By manipulating the system and through sheer persistent effort, Vincent Freeman (Ethan Hawke) soars above his genetic lacks to realize his dream of becoming an astronaut.

These stories, along with Shelley's *Frankenstein* and the many other tales about modified or wholly synthetic beings, point to two questions we must face as fiction becomes reality: the scientific one, "Can we truly modify living things and go on to build synthetic beings?" and the ethical one, which asks simply, "Even if we can, should we?"

Real synthetic life

We are still a long way from making *Blade Runner*'s replicants or a modern Frankenstein's Being, but we are now creating the science that might make this possible. One set of tools comes from advances in digital technology, robotics, neuroscience, tissue engineering, and more. These are already contributing to the design of devices and systems that compensate for bodily damage or decline, such as brain–machine interfaces that provide direct mind control for a paralyzed person to operate a computer or a person with a missing limb to manage a prosthetic replacement; and artificial bones or organs that successfully replace the natural versions. These could even someday go beyond replacing physical abilities to augmenting them and perhaps also to creating neural prosthetics for conditions such as the deterioration of memory with age. We are also beginning to see real growth in the power of artificial intelligence in applications like computer assistants that listen, interpret, and reply, and self-driving automobiles.

These abilities to create synthetic versions of the human body and brain point to the possibility of someday creating a complete synthetic being, an android, as human-seeming and as capable as Commander Data. But for now, we do not know how to even begin building a fully functioning self-aware brain; indeed we do not even understand our own consciousness, one of the great puzzles of current science. Judging by the present state of our technologies and our still limited knowledge, the construction of an android based on computer chips, artificial eyes and skin, and so on that would qualify as a true synthetic being will not happen soon; nor does this approach offer a direct way to broadly improve the human condition, because artificial augmentations or beings do not naturally reproduce but can only be manufactured.

However, another set of new tools, the rapidly developing techniques of genetic engineering and synthetic biology, offers the real possibility of creating "improved" people—whatever that may mean—and even wholly constructed humans or human variants.

This continues an old quest that began with seeking the origins of life. Probably the earliest idea about how life begins is spontaneous generation, the theory that living things can arise from inanimate dust or dead flesh. Aristotle was the great expositor of this idea, and it was widely accepted until it was unequivocally disproved in the 19th century. In 1864, the great French scientist Louis Pasteur announced experimental results showing that when microorganisms were prevented from entering a sealed environment, it remained sterile without spontaneously producing life. Instead, he concluded, life as we know it proceeds by the law of biogenesis—living beings arise only when other living beings reproduce.

But though life comes only from life, in the 19th century, chemical analysis showed that living things are made of nonliving components, and later, that life could be affected by purely chemical means. In 1899, the German researcher Jacques Loeb, working at the Marine Biological Laboratory in Woods Hole, Massachusetts, used saltwater solutions to induce sea-urchin eggs to reproduce without fertilization in the process called parthenogenesis. This achievement was overhyped in the media of the time as the "creation of life," but it supported Loeb's dream of truly controlling life. As he was quoted in an article in McClure's Magazine in 1902,

> I very early came to the belief that the forces which rule in the realm of living things are not other than those which we know in the inanimate world. Everything pointed that way...I wanted to take life in my hands and play with it. I wanted to handle it in my laboratory as I would any other chemical reaction – to start it, stop it, vary it, study it under every condition, to direct it at my will!

In 1912, Loeb's book *The Mechanistic Conception of Life* underscored these ideas and his vision of creating "a constructive or engineering biology in place of a biology that is merely analytical," that is, a science and technology that could successfully manipulate living things.

Loeb anticipated today's genetic engineering, but before that could develop, scientists had to learn more about the materials of life. Some of that knowledge came from the 19th-century development of organic chemistry, the study of the carbon-based molecules that also contain hydrogen, oxygen, and nitrogen and are ubiquitous in living things. Then in the 1930s, the new science of molecular biology began examining the molecules of life in detail, particularly the proteins that perform living functions. The powerful tool of X-ray analysis, which uses short-wavelength electromagnetic radiation to determine the atomic structure of intricate molecules, aided this development. The most significant of these efforts was the determination of the structure of deoxyribonucleic acid, DNA. When that complex biological molecule was first isolated in 1869, its function was unknown, but later there were hints that it passed on genetic information. In 1952, researchers confirmed that it did so, but exactly how was not understood.

The answer came in 1953, when the American scientist James Watson and the British scientist Francis Crick used X-ray data to show that the DNA molecule consists of a double helix, chains of smaller units that twist around each other like a double spiral staircase. These units include adenine, guanine, cytosine, and thymine (A, G, C, and T) appearing in "base pairs"—always A with T, and C with G—whose arrangement encodes genetic information. The pairing ensures that this information is copied and passed on when an organism reproduces and its cells divide. The importance of the discovery was recognized when Watson, Crick, and a third researcher, Maurice Wilkins, were awarded the Nobel Prize in Physiology or Medicine for 1962 (questions remain about whether a fourth researcher, Rosalind Franklin, has received proper credit for the important role of her X-ray data).

In establishing how heredity operates, the discovery of the structure of DNA was a major step in understanding life processes. The discovery also opened the possibility of characterizing an organism by examining its DNA or "genome," since particular sequences within the DNA—the genes—act as instructions to make the proteins that determine how an organism functions. This laid the basis for genetic manipulation. As Jacques Loeb had dreamt, DNA "blueprints" could be chemically altered to engineer the resulting lifeform. Victor Frankenstein had to stitch together body parts to build his creature, but a modern scientist could in principle modify or design an organism by changing or creating its DNA.

To understand and manipulate an organism's genome, a scientist must first determine how the base pairs of its DNA are arranged to form its genetic code, a process called "sequencing," and then find connections between specific genes and the bodily features and functions they control. Human DNA contains some three billion base pairs and over 20,000 genes, so sequencing it is a daunting task. But in 2000, Francis Collins, Director of the U. S. National Institutes of Health (NIH), and the pioneering geneticist J. Craig Venter announced that human DNA had been sequenced through the efforts they led. This was done jointly though the Human Genome Project, supported by the NIH and an international scientific consortium; and a parallel privately funded effort by the Celera Corporation, spearheaded by Venter. The full genome, published in 2003, was described by the NIH as "nature's complete blueprint for building a human being."

Progress has also been made in determining the connections between human genes and the bodily traits they define, which is essential for genetic knowledge to produce true medical benefits. This is not simple because genes and what they control may not be connected in a direct one-to-one way. A given gene may influence more than one trait; a given trait such as autism may be related to genes scattered throughout the entire genome; and environmental and other factors can influence how cells actually interpret the genetic instructions. Nevertheless, it has been possible to identify, for instance, genes that determine blood type, play a role in resistance to infection, and are linked to early onset breast cancer along with about 1800 other diseases.

In some cases, researchers now know enough about the human genome to carry out targeted genetic engineering. An early example resulted in a better way to make insulin, the protein that controls the amount of sugar in the bloodstream. Diabetics lack this natural compound and must receive it by injection. This injectable insulin used to be derived from cows or pigs; but in 1973, scientists developed a better approach using recombinant DNA, which was constructed in the lab from other DNA sequences. They isolated the part of human DNA that produces insulin and inserted it into the DNA of certain bacteria by chemical means. As the bacterial cells divided and reproduced, they produced insulin that could be harvested. Today this engineered insulin is used by millions of diabetics.

Recombinant technology that splices desirable genes into existing DNA has other applications in medicine such as producing human growth hormone, and elsewhere. In agriculture, for example, the technique has created crops like corn and soybeans modified to resist pests, disease, and herbicides and that are now widely grown. Beginning 10,000 years before the advent of modern bioscience, humanity has obtained similar desirable results through the selective breeding of plants and animals, a completely accepted method of manipulating genetic outcomes. Nevertheless, some consumers emphatically reject genetically modified "Frankenfood" as unnatural or insufficiently tested, illustrating the strong negative reactions that genetic engineering can generate even when not applied to people.

Both the beneficial and worrisome aspects of genetic engineering are amplified with the rise in the last decade of synthetic biology, an advanced approach that provides greater control over living things. Synthetic biology is so new that it lacks a universal definition, but a core element is that it is an engineering approach to manipulating biological systems. One definition is that it deals with "the design and construction of novel artificial pathways, organisms or devices, or the redesign of existing biological systems." Another formulation, made by a group of experts for a European Commission report, defines synthetic biology as

> the synthesis of complex biologically based (or inspired) systems, which display functions that do not exist in nature...at all levels of the hierarchy of biological structures – from individual molecules to whole cells, tissues and organisms.

The "systems" aspect of synthetic biology also draws on digital and computational science. In a recent review, Simon Auslander of ETH Zurich and Martin Fussenegger of the University of Basel point out that like transistors, which control the flow of electrical current and can be connected in circuits to perform digital operations, "biomolecules can control the flow of biological signals and are connected to complex circuits that organize cellular operations."

An example of synthetic biology in action is the production of the anti-malarial drug artemisinin, discovered as a plant extract by the Chinese scientist Youyou Tu, who in 2015 shared a Nobel Prize for her work. With its natural origin, the cost and availability of artemisinin suffered undesirable fluctuations. In 2004, biochemical engineer Jay Keasling at the University of California – Berkeley began developing an artificial multi-step "metabolic pathway" to make the drug cheaply and reliably. His team synthesized the gene that made an enzyme that is necessary to produce the drug, inserted the gene into yeast cells that then made artemisinin, and optimized the process for large-scale production. By mid-2015, the method had supplied 15 million doses to African countries where malaria is a serious issue and is projected to provide the drug to 1 billion people by 2025. Other applications of synthetic biology are emerging. In

2015, scientists at the J. Craig Venter Institute (JCVI, the research establishment founded by Venter) showed how to modify diatoms, a type of microalgae found in watery environments, to produce biofuels.

Despite its real or projected benefits, synthetic biology raises doubts like the concerns about Frankenfood, and inspires fears of "making monsters" and "playing God." In 2006, an article in *The Economist* magazine introducing this new science was provocatively titled "Playing demigods" and showed an image of Frankenstein's monster contemplating DNA's double helix. The article reasonably asserted that "synthetic biology needs to be monitored, but not stifled," yet also exploited fears of what would happen when the first living thing "created from scratch by the hand of man" begins to reproduce. These fears persist a decade later, as in a 2016 article from The Institution of Engineering and Technology entitled "Frankenstein redux: Is modern science making a monster?"

Apocalyptic language aside, the rapid advances in genetic manipulation force us to ask whether we can indeed radically modify people or even make artificial versions. The closest we have come so far is with a lifeform far simpler than a human, a radically re-engineered microbe. In 2010, Venter and a team of researchers at JCVI built an entire genome in the lab, made to be similar to that from a natural bacterium, *Mycoplasma mycoides.* This DNA was inserted into cells of a related bacterial strain *Mycoplasma capricolum* to replace its natural genome that had been removed, like changing a computer's operating system while leaving the hardware in place. The modified bacteria continued to function and reproduce, and following the instructions encoded in the new DNA created proteins characteristic of *Mycoplasma mycoides.* Effectively *M. capricolum* had been changed into *M. mycoides.*

According to the researchers, this experiment was carried out to learn how to manipulate DNA more effectively, not to create a synthetic being. In any case, it produced only a synthetic genome, but like Loeb's work on parthenogenesis, it was presented in the media as the creation of a complete synthetic organism. Some religious groups denounced the result, and Venter was accused of "playing God." Still this achievement, remarkable as it is, does not match the complexity of dealing with the human genome. Syn 1.0, the artificial version of the *Mycoplasma mycoides* genome, contains 901 genes, compared to the more than 20,000 in the human genome.

In practice, the brave new synthetic world contains many pitfalls in editing or making a genome, even a relatively simple one, and putting it into an organism; and even if that can be accomplished, it is still difficult to know exactly what genes control which of an organism's traits and functions. In early 2016, Venter's team at the JCVI reported on efforts to engineer a bacterium with the smallest genome and fewest genes that would support life. Two different attempts to build this minimal DNA for *M. capricolum* from scratch did not produce a viable microbe, though through trial and error the team did produce a pared-down version of Syn 1.0 with only 473 genes.

But genomic knowledge is growing fast and so is the push by entrepreneurs and investors as well as scientists and clinicians to rapidly explore, develop, and exploit the area. This heightens concerns about how, whether and when we should manipulate the human genome, and who, if anyone, should do so.

These issues were raised by researchers, clinicians, technologists, and bioethicists, led by the Nobel Laureate David Baltimore, in the leading research journal *Science* in April 2015, and at a summit conference in December 2015 convened by the U. S. National Academy of Sciences and other institutions. In both venues, the focus was on the new gene editing method called CRISPR, an acronym for "clustered regularly interspersed palindromic repeats." Named the scientific breakthrough of 2015, CRISPR makes it relatively simple to cut open a DNA sequence at a precisely targeted site and splice in a new sequence—for instance, to eliminate a sequence that is known to control a harmful disease and replace it with a corrected set of DNA instructions. In general, as the *Science* article states, this throws gene editing wide open by allowing "any researcher with knowledge of molecular biology to modify genomes, making feasible experiments that were previously difficult or impossible to conduct."

CRISPR has already shown its value, for instance, in developing therapies for cancer and AIDS. Its great potential to treat disease and even improve human capabilities has been quickly recognized by commercial interests. The profit-making possibilities for the technique have led to contentious disputes over patent rights. In early 2017, the U. S. Patents and Trademark Office decided between competing claims for its invention from the Broad Institute, an arm of MIT and Harvard that explores genetic knowledge for the treatment of disease; and the University of California. The Patents Office decided in favor of the Broad Institute, but the University of California may appeal the decision.** Despite this lingering uncertainty, in what has been called "CRISPR Inc.," an estimated billion dollars of start-up capital has poured into firms that will develop the technique for applications that promise revolutionary advances in areas from medicine to agriculture.

Real synthetic life and real ethics

There is, however, one great issue with genetic manipulation that sets it apart from the technology that could produce mechanical and digital android versions of people. Genetic technology could be used to change the DNA in the reproductive cells that carry genetic information to an organism's offspring, its so-called "germline." If this is done in sperm, fertilized eggs, or embryos, every cell in the organism is altered and the changes, for good or ill, are forever passed on to succeeding generations. This possibility greatly raises the stakes for what genetic manipulation could do to the human race and, therefore, raises serious ethical issues.

With these questions in mind, David Baltimore said at the 2015 summit that "the unthinkable has become conceivable." He and his colleagues believe it would be "irresponsible to proceed" without further evaluating the risks for humanity and gaining a "societal consensus" about how to continue. As a temporary solution, they called for a moratorium on making inheritable changes to the human genome.

But by the time the moratorium was proposed, its spirit had already been violated. Earlier in 2015, a Chinese research group became the first to openly report an attempt to alter human DNA with CRISPR. The experiment was carried out on abnormal non-viable embryos from in vitro fertilization clinics, so there was no intention to produce a baby; nor did the experiment directly raise issues according to current ethical protocols. Still, the effort was controversial and its outcome disturbing. The researchers hoped to produce embryos with a precisely altered gene in every cell but met utter failure. Every one of the 85 embryos either died, lacked the altered gene, exhibited a mixture of altered and unaltered cells, or suffered damaged DNA.

This failed attempt and a similar effort proposed in the U. K. motivated the December 2015 summit conference and the resulting recommendation for a moratorium. But in February 2017, the National Academy of Sciences and the National Academy of Medicine—leading commenters on and arbiters of scientific and medical policy in the U. S.—issued a lengthy report, "Human Genome Editing: Science, Ethics, and Governance," which further examined the ethical questions. An international group of nearly two dozen scientists and physicians, bioethicists, and representatives of the legal, cultural, and commercial worlds reviewed the tools available for genetic editing and the consequences of such alterations.

The report notes first that as a society and as a matter of public policy, we have not really begun to address the many human implications of genetic technology. One concern with heritable genome editing is to find the balance between the benefits that would come to individuals such as parents and children, and the harms that could affect an entire society and culture. "This is a complicated ethical analysis," says the report,

> because the individual benefits and risks are more immediate and concrete, whereas concerns about cultural effects are necessarily more diffuse…and because any cultural changes resulting from a new technology take time to develop. (*Human Genome Editing*, 91).

Nevertheless, we must find a way to weigh the individual benefits against the wider costs. These include unintended genetic consequences, and the impact on the ideal of a "level playing field" for all the members of a society. In the former instance, if people could be physically and mentally enhanced via gene editing as in the film *Gattaca*, that would change human DNA with unknown long-term outcomes for descendants. In the latter, if enhancements were limited to those who could afford them, that would only make worse the inequalities in a society where lack of equal opportunity is already an issue. Even if gene editing were used not to enhance people but only to eliminate disease, that too could become available only to the "haves" in our society within the present system of distributing medical care. Overall, the potential split between those who would benefit from gene editing and those who would not or would even be relatively diminished by it carries an uncomfortable whiff of eugenics.

For reasons such as these, the report recommends that genome editing should not proceed without more consideration of the issues; but to the surprise of many observers, the report does not support a full moratorium. Instead, it proposes that after more studies of risks and benefits, clinical trials of embryo editing might be allowed for couples if both have a serious genetic disease that might be passed on to a child for whom embryo editing is "the last reasonable option." Predictably, illustrating the weight of the unresolved ethical issues, this engendered a diversity of responses from the scientific community. They ranged from approval of the limited clinical use to dismay that the report seemingly supports any use whatsoever of genetic editing before a full public discussion and consensus have taken place.

Meanwhile experiments continue in human genetic editing when they do not violate the restriction that the genetic changes cannot be heritable. In early 2017, a team of Chinese scientists reported using CRISPR to correct genetic mutations in six human embryos. Unlike the embryos used in the earlier failed Chinese effort in 2015, these were normal except that they were immature and so could not develop into babies. The researchers were able to correct the mutation, which would have led to a hereditary disease, in all the cells of one embryo and in some cells in two others—a limited but better success rate than in the 2015 attempt.

It has also become apparent that scientists are ready to go beyond gene editing tools like CRISPR with still bolder approaches to modifying the human genome. In 2016, 25 academic scientists and figures in the biotechnology industry, led by Jef Boeke at New York University and George Church at Harvard, proposed the Human Genome Project–Write (HGP-Write). The original Human Genome Project had sequenced or "read" the human genome. HGP-Write aims to create from scratch or "write" the entire human genome with its 3 billion base pairs, and other similarly huge genomes, within a decade at a cost of billions.

Considering that so far the only lab-built genomes are the small bacterial ones from JCVI and a portion of the yeast genome made by Boeke, this is a remarkably audacious proposal. It met mixed responses from the scientific community. Some researchers called it "brilliant" with "unlimited potential"; others questioned its scientific rationale and whether existing technology can really design and build enormous genomes. But the project's originators think this approach is essential if we are to make genetic change on a big scale. "Editing [of genes] doesn't scale very well," said Church, adding, "When you have to make changes to every gene in the genome it may be more efficient to do it in large chunks."

Though the founders of HGP–Write pointed to its scientific and medical benefits, observers raised ethical concerns. For instance, Drew Endy, a bioengineer at Stanford, and Laurie Zoloth, a religion professor at Northwestern University, said that questions such as "whether and under what circumstances we should make such technologies real" should have been broadly addressed before the project began and not decided by a small group without seeking wide consensus.

HGP-Write would be managed by a nonprofit operation, the Center of Excellence for Engineering Biology. But observers are asking whether commercial support for the effort would entail intellectual-property restrictions that limit open access to its research, nor is it clear that the Federal research establishment would support the concept and fund the project. Tellingly, NIH Director Francis Collins was not immediately enthusiastic. The NIH encourages research in DNA synthesis, he said, but "has not considered the time to be right for funding a large-scale production-oriented" project like this, adding that such projects "immediately raise numerous ethical and philosophical red flags."

In response to the ethical questions, the project's founders were quick to discount fears that it would directly affect the human germline or lead to the creation of new kinds of humans. Jef Boeke said that any synthetic products from HGP-Write would be engineered to make reproduction impossible, adding, "We're not trying to make an army of clones or start a new era of eugenics. That is not the plan."***

This assurance and the proposed moratorium on modifying human DNA except for limited clinical use, if followed, would seem to allay fears that HGP-Write, CRISPR, and synthetic biology will soon lead to a world like that in the film *Gattaca,* dominated by genetic destiny and a kind of new scientific eugenics. The scientific potential certainly remains for improved or completely new kinds of beings to be created using these methods, but Frankenstein-like moments of creation are still far distant, according to two leading scientists involved in these efforts. After the failure in 2016 to design and make a minimal bacterial genome at JCVI, Craig Venter summarized the state of the art when he said, "Our current knowledge of biology is not sufficient to sit down and design a living organism and build it." And responding to the HGP-Write proposal to construct the entire human genome, NIH Director Collins said "whole-genome, whole-organism synthesis projects extend far beyond current scientific capabilities."

This does not mean that synthetic life will never come, because the pressures to continue exploring and using the relevant science and technology will continue; but if it does, it will not appear as living goo crawling out of a test tube or a Frankensteinian creature suddenly rising up from a laboratory bench. Rather it will be the result of modifying humans or other organisms over succeeding generations. This long-term approach to Victor Frankenstein's dream of "infusing life" may give us time to understand what we are doing and why we are doing it. If so, we may hope to someday make and deal with synthetic lifeforms without the "breathless horror and disgust" that Victor Frankenstein felt, but with the knowledge that we have achieved something good for humanity and for the new beings themselves.

---

*In 2018, Chinese scientist He Jiankui was found to have gene-edited human babies. He was widely criticized, his research was suspended, and he was sentenced to a prison term.

**As of September 2020, this decision was still being adjudicated.

***For the current status of HGP-Write, see https://engineeringbiologycenter.org/about/

References

Asimov, Isaac. *I, Robot*. New York: Gnome Press, 1950.

Auslander, Simon and Martin Fussenegger. From gene switches to mammalian designer cells: Present and future prospects. *Trends in Biotechnology* 2013, **31**, 3, pp. 155–168.

Ball, Philip. Man made: A history of synthetic life. *Distillations*, 2016, **2**, 1, pp. 14–23. https://www.chemheritage.org/distillations/magazine/man-made-a-history-of-synthetic-life.

Baltimore, David et al. A prudent path forward for genomic engineering and germline gene modification. *Science* 2015, **348**, 6230, pp. 36–38. http://science.sciencemag.org/content/348/6230/36.

Boeke, Jef D. The Genome Project–Write. *Science* 2016, **353**, 6295, pp. 126–127. http://science.sciencemag.org/content/early/2016/06/03/science.aaf6850.

Callaway, Ewen. Plan to synthesize human genome triggers mixed response. *Nature* 2016, **534**, 163. http://www.nature.com/news/plan-to-synthesize-human-genome-triggers-mixed-response-1.20028.

Cohen, Jon. The Birth of CRISPR Inc. *Science* 2017, **355**, 6326, pp. 680–684. http://science.sciencemag.org/content/355/6326/680.

Dick. Philip K., *Do Androids Dream of Electric Sheep?* New York: Ballantine Books, 1996.

European Commission, Synthetic Biology. Applying Engineering to Biology. 2005. http://www.synbiosafe.eu/uploads/pdf/EU-highlevel-syntheticbiology.pdf.

Explore More: Genetic Engineering. Recombinant DNA: Example Using Insulin. Iowa Public Television, http://www.iptv.org/exploremore/ge/what/insulin.cfm.

Fell, Jade. Frankenstein redux: Is modern science making a monster? *Engineering and Technology*, 2016. https://eandt.theiet.org/content/articles/2016/06/frankenstein-redux-is-modern-science-making-a-monster.

Freud, Sigmund. The Uncanny. http://web.mit.edu/allanmc/www/freud1.pdf.

Gibson, Daniel G., et al. Creation of a bacterial cell controlled by a chemically synthesized genome. *Science* 2010, **329**, 5987, pp. 52–56. http://science.sciencemag.org/content/329/5987/52.

Harmon, Amy. Human gene editing receives science panel's support. *The New York Times*, Feb. 14, 2017. https://www.nytimes.com/2017/02/14/health/human-gene-editing-panel.html.

Hotz, Robert Lee. Scientists create synthetic organism. *The Wall Street Journal*, May 21, 2010. http://www.wsj.com/articles/SB10001424052748703559004575256470152341984.

Huxley, Aldous. *Brave New World*. New York: Alfred A. Knopf, 2013.

J. Craig Venter Institute. http://www.jcvi.org/cms/home/.

Kaiser, Jocelyn. U.S. panel gives yellow light to human embryo editing. *Science* 2017. http://www.sciencemag.org/news/2017/02/us-panel-gives-yellow-light-human-embryo-editing.

Karas, Bogumil J. et al. Designer diatom episomes delivered by bacterial conjugation. *Nature Communications* 2015, **6**, 6925. http://www.nature.com/articles/ncomms7925.

Karel, Čapek. *R. U. R. (Rossum's Universal Robots)*. H. Milford, London; New York: Oxford University Press; 1925. http://preprints.readingroo.ms/RUR/rur.pdf.

Katsnelson, Alla. Researches start up cell with synthetic genome. *Nature* 2010. http://www.nature.com/news/2010/100520/full/news.2010.253.html.

Kozubek, Jim. *Modern Prometheus: Editing the Human Genome with CRISPR-CAS9*. Cambridge and New York: Cambridge University Press, 2016.

Ledford, Heidi. Court rules on CRISPR. *Nature* 2017, **542**, 401. https://iatranshumanisme.files.wordpress.com/2017/02/nature-2017-21502.pdf.

Le Page, Michael. First results of CRISPR gene editing of normal embryos released. *New Scientist Daily News*, March 9, 2017. https://www.newscientist.com/article/2123973-first-results-of-crispr-gene-editing-of-normal-embryos-released.

Liang, Puping et al. CRISPR/Cas9-mediated gene editing in human tripronuclear zygotes. *Protein & Cell* 2015, **6**, 5, pp. 363–372.

Lussier, Germain. NASA says "2012" is most absurd sci-fi movie ever; "Gattaca" most plausible. Film, http://www.slashfilm.com/nasa-2012-absurd-scifi-movie-gattaca-plausible/, Jan. 6, 2011.

Mirchandani, Aneela. The original Frankenfoods: Origins of our fear of genetic engineering. Genetic Literacy Project, Feb. 10, 2015. https://www.geneticliteracyproject.org/2015/02/10/the-original-frankenfoods/.

Mori, Masahiro. The Uncanny valley. http://spectrum.ieee.org/automaton/ robotics/humanoids/the-uncanny-valley.

National Academy of Sciences and National Academy of Medicine. *Human Genome Editing: Science, Ethics, and Governance*. Washington, DC: The National Academies Press, 2017. https://www.nap.edu/ catalog/24623/human-genome-editing-science-ethics-and-governance.

National Library of Medicine, NIH. How did they make insulin from recombinant DNA? https://www.nlm.nih.gov/exhibition/ fromdnatobeer/exhibition-interactive/recombinant-DNA/ recombinant-dna-technology-alternative.html.

NIH National Human Genome Research Institute. All About the Human Genome Project. https://www.genome.gov/10001772/.

NIH Research Portfolio Online Reporting Tools (RePORT). Human Genome Project. https://report.nih.gov/nihfactsheets/ViewFactSheet. aspx?csid=45\.

Nobelprize.org, The Nobel Prize in Physiology or Medicine 1962. https:// www.nobelprize.org/nobel_prizes/medicine/laureates/1962/.

Nobelprize.org. The Nobel Prize in Physiology or Medicine 2015. https:// www.nobelprize.org/nobel_prizes/medicine/laureates/2015/.

Norrgard, Karen. Human testing, the eugenics movement, and IRBs. *Nature Education* 2008, **1**, 1, pp. 170.

http://www.nature.com/scitable/topicpage/human-testing-the-eugenics-movement-and-irbs-724.

Pauly, Philip J. The invention of artificial parthenogenesis. Chapter 5 in *Controlling Life: Jacques Loeb and the Engineering Ideal in Biology*, Oxford University Press, NY. pp. 93–117. 1987. http://10e.devbio. com/article.php?id=72.

Perkowitz, Sidney. *Digital People*. Washington, DC: Joseph Henry Press, 2004.

Perkowitz, Sidney. Digital people in manufacturing: Making them and using them. *The Bridge* 2005, **35**, 1, pp. 21–25.

Perkowitz, Sidney. Resistance is unnecessary: Accepting the cyborg in our midst. *Literal* 2010, **19**, pp. 26–27.

Perkowitz, Sidney. Cuerpo y Mente Unidos Por un Chip (Body and Mind Joined by a Chip). *Quo* 2014, pp. 40–44.

Perkowitz, Sidney. Removing humans from the AI loop: Should we panic? *Los Angeles Review of Books* 2016.

Perkowitz, Sidney. How to understand the resurgence of eugenics. *JSTOR Daily*, April 5, 2017. https://daily.jstor.org/how-to-understand-the-resurgence-of-eugenics/.

Plumer, Brad. Scientists can now genetically engineer humans. A big new report asks whether we should. *Vox*, February 15, 2017. http://www.vox.com/science-and-health/2017/2/15/14613878/national-academy-genome-editing-humans.

Pollack, Andrew. Scientists announce HGP-Write, Project to synthesize the human genome. *The New York Times,* June 2, 2016. http://www.nytimes.com/2016/06/03/science/human-genome-project-write-synthetic-dna.html?mwrsm=Email.

Roosth, Sophia. *Synthetic: How Life Got Made*. Chicago: University of Chicago Press, 2017.

Sample, Ian. Craig Venter creates synthetic life form. *The Guardian*, May 20, 2010. https://www.theguardian.com/science/2010/may/20/craig-venter-synthetic-life-form.

Science News Staff. And *Science's* 2015 Breakthrough of the Year is.... *Science* 2015. http://www.sciencemag.org/news/2015/12/and-science-s-breakthrough-year.

Service, Robert F. Synthetic microbe lives with fewer than 500 genes. *Science* 2016, http://www.sciencemag.org/news/2016/03/synthetic-microbe-lives-less-500-genes.

Shelley, Mary Wollstonecraft. *Frankenstein* (1818). https://archive.org/stream/Frankenstein1818Edition/frank-a5_djvu.txt.

Shelley, Mary Wollstonecraft. *Frankenstein* (1831). http://tcpl.org/community-read/Frankenstein/mary1831.pdf.

Shelley, Mary Wollstonecraft. *Frankenstein: Complete, Authoritative Text with Biographical, Historical, and Cultural Contexts, Critical History, and Essays from Contemporary Critical Perspectives*, edited by Johanna M. Smith. Boston: Bedford/St. Martin's, 2000.

Snyder, Carl. Bordering the mysteries of life and mind. *McClure's Magazine* 1902, **18**, pp. 386–396. http://tinyurl.com/j87hscc.

Synthetic Biology Project. Synthetic Biology 101. http://www.synbioproject.org/topics/synbio101/definition/.

The Center of Excellence for Engineering Biology. Introducing GP-write. http://engineeringbiologycenter.org/.

The Economist. Playing demigods. Aug. 31, 2006. http://www.economist.com/node/7854771.

Tracinski, Rob. The future of human augmentation and performance enhancement. *Real Clear Science*, April 4, 2017. http://www.realclearscience.com/articles/2017/04/04/the_future_of_human_augmentation_and_performance_enhancement.html.

Travis, John. Germline editing dominates DNA summit. *Science* 2015, **350**, 6266, pp. 1299–1300. http://science.sciencemag.org/content/350/6266/1299.

Twist Bioscience. Synthetic biology assures global access to a vital Nobel Prize winning malaria medication. Nov. 23, 2015. https://www.twistbioscience.com/dr-jay-keasling/.

Wade, Nicholas. Scientists seek moratorium on edits to human genome that could be inherited. *The New York Times,* Dec. 3, 2015., http://www.nytimes.com/2015/12/04/science/crispr-cas9-human-genome-editing-moratorium.html.

Watson, James D. *The Double Helix.* New York: Scribner, 1998.

Filmography and videography

Blade Runner (Ridley Scott, 1982).

Forbidden Planet (Fred M. Wilcox, 1956)

Frankenstein (James Whaley, 1931).

Gattaca (Andrew Niccol, 1997).

Metropolis (Fritz Lang, 1927).

Star Trek: the Next Generation (Television series, 1987–1994).

The Day the Earth Stood Still (Robert Wise, 1951).

Intelligent Machines: It's More Than Just Intelligence

Science fiction stories—whether set on Earth or in a distant galaxy—are likely to feature intelligent machines, often as main characters. They show up as disembodied artificial intelligences like the HAL 9000 computer that operates a Jupiter-bound spaceship in *2001: A Space Odyssey* (1968); with mechanical bodies like the hulking robot Gort in *The Day the Earth Stood Still* (1951, 2008) or the elegant and charming robot C-3PO in *Star Wars* (1977); or as human-looking androids like the T-800 unit in *The Terminator* (1984), Commander Data in *Star Trek: The Next Generation* (1987–1994), and the replicants in *Blade Runner* (1982) and *Blade Runner 2049* (2017).

These machines appear under different names—robots, synthetic beings, androids, AIs, droids, cyborgs, replicants—but whatever their name and form, they're smart, autonomous, and physically capable, though sometimes limited by built-in constraints. They are often scary but can also have an appealing side, and they are always intriguing.

Why this fascination with watching artificial versions of our minds and bodies? Maybe we want to see technology pushed to the point where we can feel like gods, with the power to design and create living or semi-living beings. Behind that may be a secret human longing: If we can do that, maybe someday we will know how to improve ourselves. Or maybe it's the desire to see ourselves on screen, but indirectly, through our own creations. That gives us a vantage point from which to honestly contemplate our human sins and virtues. But our goals may not be that lofty. We might just like imagining a world where mechanical servants do the things we'd rather not do ourselves, or wait on us with inhuman perfection.

These reasons must lie deep in the human psyche because long before we could make intelligent machines, they were part of human fantasy. The idea of a smart metal robot goes all the way back to a story in Greek mythology about Talos, a man-shaped creation made of bronze that patrolled the island of Crete and defended it by lobbing rocks at approaching ships.

Talos was the first of a long line of robots that have been put to use by humans. Their low status was clearly defined in the

1921 stage play *R. U. R.* (*Rossum's Universal Robots*) by the Czech writer Karel Čapek (followed by a Russian film version in 1935). It featured human-looking artificial laborers made in a factory and called "robots," a word that in Czech meant "forced labor" or "slave." These creations eventually become self-aware, develop emotions, and grow to deeply resent their position. They rebel and wipe out humanity, but the film leaves a ray of hope when a male and female robot discover love and go off to find a new—and maybe better—race.

The theme of robots as slaves has continued since, from the classic Fritz Lang film *Metropolis* (1927) where a scientist creates the eerily feminine robot Maria as the first of a horde of robot workers, to the replicants in director Ridley Scott's *Blade Runner* who are made to work for humanity as it settles distant planets. The latter film's synthetic beings, also known by the pejorative "skinjobs," seem very nearly human but are considered disposable, with lifetimes of only 4 years. Led by replicant Roy Batty (Rutger Hauer), a group of them murders a human spaceship crew and illegally returns to Earth hoping to get their lives extended.

Special agent Rick Deckard (Harrison Ford), the "Blade Runner" of the title, is assigned to hunt down the rebel replicants, destroying them all except for Batty. In the final scene, speaking dialogue written by actor Rutger Hauer himself, Batty makes a poetic speech about the experiences and memories that will be lost at his death. He then expires, his allotted time having come to an end. The scene has become a standout moment in science fiction because the powerful imagery in Batty's speech and Hauer's subtle portrayal of a synthetic being becoming human (and perhaps more than human) transcend stereotypes about robots.

This depiction of machines that initially have no claim on our sympathies but later become human enough to elicit our empathy is also explored in other science fiction films. In director Stanley Kubrick's *2001*, the artificial intelligence HAL kills a human crew member to preserve itself—yet we feel compassion when Commander Bowman (Keir Dullea) pulls out the AI's memory units and we watch this capable mind regress to that of a five-year-old child singing "Daisy, Daisy, give me your answer do." In *Star Trek: The Next Generation*, the android Commander Data (Brent Spiner) is accepted and respected by his human Starfleet shipmates. He is stronger and smarter than humans, but still he aspires to become more human. His naive and earnest efforts to learn about emotions and deal with a pet cat have made Data a lovable character in the series, and those efforts are a compliment to us, his creators. Every parent is flattered by a child who wants to be more like him or her.

Other fictional machine intelligences are not at all lovable. The T-800 unit in James Cameron's *The Terminator* (played by Arnold Schwarzenegger) looks like a person, albeit a stone-faced one. But his synthetic outer layer is only a disguise for a machine with a single implanted directive: Find and kill the woman whose unborn son will lead the future resistance against Skynet, the self-aware computer network that wants to wipe out humanity. In *Battlestar Galactica* (2004–2009), the Cylons (Cybernetic Lifeform Nodes) who appear in both humanoid and machine-like forms have destroyed most of humanity, which they consider a flawed race, and want to finish off the last survivors.

Whether the synthetic beings are shown as worthy of human concern or as enemies of humanity, these stories do something that great science fiction always does (besides entertain): They map out possible futures and allow us to imagine where new technology may take us before the technology actually arrives. When *Blade Runner* was released in 1982, robotics, AI, and genetic engineering—the sciences that would actually lead to synthetic beings of one kind or another—were in their infancy. Yet *Blade Runner* foresaw issues we must seriously consider 35 years later, after the technology has advanced enormously. Newer films such as Steven Spielberg's *A.I. Artificial Intelligence* (2001) and director Alex Garland's acclaimed android thriller *Ex Machina* (2014) have continued to address these concerns.

At this time, we do not have any real synthetic beings as capable as the movie versions. No robots and androids made today look convincingly human or move like people. Nor does current AI demonstrate broad intelligence like we see onscreen, though some enthusiasts such as the futurist Ray Kurzweil believe we are on the verge of creating human-level machine intelligence. Others, like the British roboticist Murray Shanahan, agree that we will produce advanced AIs but not in the near future. Shanahan, who was scientific adviser for *Ex Machina*, thinks that current digital technology might be able to simulate the 70 million neurons in a mouse brain. But that amount is less than 0.1 percent of the 80 billion neurons in the human brain, so creating human-level general intelligence such as that of *Blade Runner's* replicants remains a long way off.

Still, AI and robots are entering our world at an increasing pace, and we have to learn to live with them. Back in 1950, science fiction writer Isaac Asimov gave us clues to how the interaction might work in his book *I, Robot*, which established the Three Laws of Robotics: (1) A robot may not injure a human being or, through inaction, allow a human being to come to harm; (2) A robot must obey orders given to it by human beings, except where such orders would conflict with the First Law; and (3) A robot must protect its own existence as long as such protection does not conflict with the First or Second Law. Later Asimov added a "Zeroth Law" that preceded the others: A robot may not harm humanity, or, by inaction, allow humanity to come to harm.

I, Robot, the 2004 Will Smith film loosely based on Asimov's writings, prominently incorporated the Three Laws in its depiction of a 2035 civilization where robots are widely used and trusted to help humanity. But even these apparently bulletproof guides can be violated. In the story, one robot is found to have murdered a person under special circumstances that allowed it to evade the First Law. Even worse, a higher-order AI that controls the robots develops its own interpretation of the Three Laws and deduces its own Zeroth Law: that its highest duty is to all humanity. The AI orders all the existing robots to take control of humanity in order to save us from ourselves. The resulting robot revolution is only barely stopped at the end of the film.

The problem is not just that rigid, implanted rules like the Three (or Four) Laws could be interpreted in unexpected ways by a truly smart AI. It is also that such directives are too inflexible to deal with real ethical questions. These could be coming up sooner than you think—for instance, on the battlefield. The U. S. does not have humanoid Terminator units that we send out on missions to kill, but we are on the way to developing autonomous weapons that could make deadly decisions in warfare.

The morality of AI-based weapons that can kill was discussed at the First International Symposium on Roboethics in 2004. Now it is being considered at the UN, as the U. S. and other nations develop military AI that would let them wage war with fewer human soldiers on the battlefield. For several years, the U. S. has used armed semiautonomous drones to find enemy combatants, with the final decision to fire at them made by human controllers thousands of miles away. The next step will involve fully autonomous weapons that decide on their own what to target and when to shoot. The potential for negative outcomes in such a program was convincingly shown in the film *Robocop* (1987), where an aggressively autonomous police robot, ED-209, kills an innocent bystander.

But according to U. S. Deputy Secretary of Defense Robert Work, full autonomy is not part of The Pentagon's plan. Instead, he said recently, the idea is to "keep humans in the decision cycle to use lethal force.... Will we ever build a robot that is completely autonomous that will exert lethal force? I think the answer to that is no." Others think the rapid development of AI will lead to an escalating AI arms race. Yet we do not know how to build an ethical war robot that can discriminate between friend and enemy or combatant and noncombatant—a complex judgment far beyond the simplicity of Asimov's First Law. With this in mind, in 2015, over 3000 robotics and AI researchers signed an open letter asking for "a ban on offensive autonomous weapons beyond meaningful human control."

To complement a moral code for robots more sophisticated than the Three Laws, we would also need a moral code for humans that will ensure we treat robots as more than slaves if they ever become as sentient as Roy Batty. There are indications that society is beginning to recognize this possibility, although the issue is not without controversy and unexpected spin-offs. In October 2017, the Kingdom of Saudi Arabia granted citizenship to Sophia, a feminine humanoid robot with some ability to carry on a conversation. Seen mostly as a public relations ploy from a country that wants to appear tech friendly and has an unfortunate record on women's rights, this action nevertheless focused attention on how rights for synthetic others should really grow out of our own human rights.

Meanwhile, the European Union is seriously considering whether there will eventually be a need to assign personhood to high-level robots and AIs. This would not make them citizens with civil rights but, like corporate personhood, would provide a legal basis for allocating blame. For instance, if the AI in a self-driving car were to make a bad decision and injure a pedestrian, who would be responsible? The autonomous AI itself, its human designers and programmers, or the corporation that put the whole package on the road? Questions like these are early steps in developing a moral stance toward intelligent machines.

If we humans ever work out how to interact with these new artificial beings, it will be science fiction stories about intelligent machines that helped us get there. By exploring the boundary between machine and person, these stories show that we have to reckon with morality as well as intelligence when dealing with our own creations. The stories also remind us that the interactions may not always be friendly.

Scientists like Stephen Hawking have issued warnings about the downside of embracing machine intelligence without sufficient forethought. We should heed these comments, but it takes the emotional punch of a murderous HAL 9000, an implacable Terminator, or a nearly human Roy Batty to really get our attention. Their stories might just be our best reminders to be wary as we head into the AI future.

Confronting the Wall

Reviews of the films *Best of Luck with the Wall, Sleep Dealer,* and *La Que Sueña (She Who Dreams)* (These films can be seen at the Labocine site, https://www.labocine.com.)

Movies can provide pure entertainment and escape and can also bring our attention to real and important issues. That's what documentaries and much of independent filmmaking aim to do, but Hollywood's feature films can also reflect real issues as a theme or more subtly, as a true-life background woven into a made-up story. At its best, one film genre in particular, science fiction, goes beyond the present moment when it puts us on notice about significant trends in science and their future effects.

Many documentaries, feature films, and science fiction stories have reflected the reality of the atomic bomb and the nuclear era and their consequences, and are doing the same for genetic science. Now another issue important to the U. S. (and the rest of the world) is receiving onscreen attention: that is immigration, its significance heightened by President Donald Trump's proposal to build a wall between the U. S. and Mexico, and by the fate of the "Dreamers," the undocumented immigrants brought to the U. S. as children.

Three films in the Labocine archive grapple with these concerns. They use satellite technology to bring home the actuality of building a border wall, show how technology might affect immigration and immigrants, and present a Latina bioscientist who, in reality or in imagination, transforms herself into a creature with greater freedom than she feels in the U. S.

The documentary *Best of Luck with the Wall* (2017, U. S., Josh Begley) gives a visceral sense of what a huge project a border wall would be. The film begins with a simple onscreen statement that sets its scale: "The U. S.–Mexico border is 1,954 miles long." Then with no further captions, characters, or dialogue but with music, it visually tracks every mile of the border from the Pacific Ocean and San Diego all the way to its eastern end at the Gulf of Mexico.

Begley pieced together 200,000 Google Maps satellite images to create this survey. It would take three hours to fly from end to end of the border by commercial jet; but the film covers the border's every twist and turn (much of it follows the serpentine path of the Rio Grande) in just six minutes. At an effective speed of nearly

20,000 mph, you see what at first seem abstract patterns; yet as your cognition kicks in, you realize that what is flashing by is birds-eye views of cities, towns, and highways; farming areas, arid deserts, and rugged mountains, ending at the mouth of the river and a sandy peninsula in the Gulf of Mexico.

The film gives a direct sense of the size and variety of terrain that a wall would have to traverse by means of major feats of engineering. Begley realizes the limits of compressing all this into a brief experience, noting that "You can never see the entire southern border in six minutes." Then he adds what is really the point of his work, "But even in that small [time frame] perhaps it's more than what one sees when reading a headline about building a wall."

Sleep Dealer (2008, U. S. and Mexico, Alex Rivera) is a science fiction vision of what a U. S.–Mexico wall would mean for both countries and their people. A recipient of a Sundance award and other honors, the film portrays a near-future world where the U. S. has sealed itself off from Mexico and where corporations have amassed a great deal of power. Memo (Luis Fernando Peña) is a young Mexican whose father, a farmer, does what he must to obtain water after the Del Rio corporation builds a dam to block the local river and sells its water. When he is killed as an "aqua terrorist" by a remotely controlled drone gunship, Memo flees to Tijuana.

There he finds that though he cannot cross the border, he can work in a hi-tech sweatshop where workers are "jacked in;" using implanted neural nodes, they remotely operate machinery across the border. This provides cheap labor for a U. S. that, as Memo's foreman says, "wants our work but doesn't want our workers." Memo is soon remotely welding a skyscraper structure in San Diego and earning money to support his family, though he faces the possibility that the neural links can go bad and seriously injure him, as happens to other workers.

Jacking in also allows emotions to be retrieved and shared. Memo becomes involved with Luz (Leonor Varela), a young writer who uses her nodes to let paid subscribers experience what she feels, but angers Memo by sharing his story without his knowledge. This has an unexpected outcome: The story inspires the drone pilot who killed Memo's father to use the drone's firepower to blow holes in the Del Rio dam so the water can again flow. Now knowing that his town will survive, Memo abandons cyberwork and returns to farming.

A decade later, we see that *Sleep Dealer* was remarkably prophetic. Its director and co-writer Alex Rivera has said that it was the outsourcing of jobs from the U. S. that inspired the film, a trend that has only grown. Though the U. S. benefits from the labor of immigrants, today it is clear that certain elements in America, from white nationalists to government officials, would rather not have the immigrants themselves, just as the film expresses. Corporate power has grown since 2008 along with closely linked levels of economic inequality. Remotely piloted armed drones have become integral to the U. S. military arsenal, and neural control of devices is now a real technology of brain–machine interfaces that control prosthetic limbs.

La Que Sueña (*She Who Dreams*) (2017, USA, Alexis Gambis) begins as we see a young Latina woman (Stephanie De Latour) walking in New York and hearing the news that President Trump is ending the Dreamer's protections against deportation. Then we learn that she is a scientist who studies butterflies. As she examines a brightly colored wing under the microscope, small patches of her own skin take on a similar coloration and structure of tiny scales.

In the next scene she has completely transformed into a butterfly-woman with bold black and orange patterns that cover her face, arms, and a set of wings. As she walks around the city, climbs a fence, and gazes at the East River, she explains how before the change she felt limited, isolated, paralyzed, but now she finally feels "like myself" with an essence that makes her proclaim "I am a Monarch;" that is, a Monarch butterfly, which displays just these patterns.

How did she make the conversion? As a scientist, she speaks of genes and their expression in proteins so it may be the result of genetic engineering; or perhaps we are simply seeing her own wishful dream. Monarch butterflies perform vast southerly migrations that can bring up to 300 million of them to specially designated reserves in Mexico. Like the butterfly-woman climbing the fence, their ability to transcend any U. S.–Mexico wall and reach sanctuary makes Monarchs true symbols of the freedom and acceptance the Dreamers dream about.

From the hard physical reality of our southern border, to predictions about immigrants and virtual work, to the emotions of a Dreamer, these three films extend the national conversation about immigration and its human toll.

Bad Blood, Worse Ethics

Review of *Bad Blood: Secrets and Lies in a Silicon Valley Startup,* John Carreyrou (Knopf, 2018).

Next to having your doctor ask you to say "aah," having your blood examined is one of the most common diagnostic rituals. Some specific blood tests, such as that for diabetes, need just a drop from a finger stick, but broader evaluations can require a teaspoon or more of blood drawn through a needle inserted into a vein in your arm.

No one enjoys the process and some truly fear it. It is worth the pain only because blood tests can reveal conditions you and your doctor need to know about, from your body's chemical balance to signs of disease. Besides the distress, there is the cost. One recent U. S. study found a median cost of $100 for a basic blood test, with much higher costs for more sophisticated analyses. This adds up to a global blood-testing market in the tens of billions of dollars.

So when 19-year-old Elizabeth Holmes dropped out of Stanford in 2003 to realize her vision of less painful, faster, and cheaper blood testing, she quickly found eager investors. An admirer of Steve Jobs, she pitched her startup company Theranos (a portmanteau of "therapy" and "diagnosis") as the "iPod of health care." But instead it became the Enron of health care, a fount of corporate deceit that finally led to a federal criminal indictment of Holmes for fraud. She now awaits trial.[*]

Bad Blood: Secrets and Lies in a Silicon Valley Startup is John Carreyrou's gripping story of how Holmes's great idea led to Silicon Valley stardom and then into an ethical quagmire. Carreyrou is the *Wall Street Journal* reporter who first revealed that Theranos was not actually achieving what Holmes claimed, though her company had been valued at $9 billion. Her net worth neared $5 billion, and her deals with Walgreens and Safeway could have put her technology into thousands of stores, thus measuring the health of millions, until Carreyrou showed that the whole impressive edifice rested on lies.

It didn't start that way. Impressed by her creativity and drive, Holmes's faculty mentor at Stanford, writes Carreyrou, told her to "go out and pursue her dream." That required advanced technology, whereas Holmes's scientific background consisted of a year at Stanford and an internship in a medical testing lab. Nevertheless, she conceived and patented the TheraPatch. Affixed to a patient's arm, it would take blood painlessly through tiny needles, analyze the sample, and deliver an appropriate drug dosage. Her idea was good enough to raise $6 million from investors by the end of 2004, but it soon became clear that developing the patch was not feasible.

Holmes didn't quit. Her next idea was to have a patient prick a finger and put a drop of blood into a cartridge the size of a credit card but thicker. This would go into a "reader," where pumps propelled the blood through a filter to hold back the red and white cells; the pumps then pushed the remaining liquid plasma into wells, where chemical reactions would provide the data to evaluate the sample. The results would quickly be sent wirelessly to the patient's doctor. Compact and easy to use, the device could be kept in a person's home.

In 2006, Holmes hired Edmond Ku, a Silicon Valley engineer known for solving hard problems, to turn a sketchy prototype of a Theranos 1.0 card and reader into a real product. Running a tiny volume of fluid through minute channels and into wells containing test reagents was a huge challenge in microfluidics, hardly a Silicon Valley field of expertise. As Carreyrou describes it:

> All these fluids needed to flow through the cartridge in a meticulously choreographed sequence, so the cartridge contained little valves that opened and shut at precise intervals. Ed and his engineers tinkered with the design and the timing of the valves and the speed at which the various fluids were pumped through the cartridge.

But Ku never did get the system to perform reliably. Holmes was unhappy with his progress and insisted that his engineers work around the clock. Ku protested that this would only burn them out. According to Carreyrou, Holmes retorted, "I don't care. We can change people in and out. The company is all that matters." Finally, she hired a second competing engineering team, sidelining Ku. (Later, she fired him.) She also pushed the unproven Theranos 1.0 into clinical testing before it was ready. In 2007, she persuaded the Pfizer drug company to try it at an oncology clinic in Tennessee. Ku fiddled with the device to get it working well enough to draw blood from two patients, but he was troubled by the use of this imperfect machine on actual cancer patients.

Meanwhile, the second team jettisoned microfluidics, instead building a robotic arm that replicated what a human lab tech would do by taking a blood sample from a cartridge, processing it, and mixing it with test reagents. Holmes dubbed this relatively clunky device the "Edison" after the great inventor and immediately started showing off a prototype. Unease about the cancer test, however, had spread, and some employees wondered whether even the new Edison was reliable enough to use on patients. As Carreyrou relates, Holmes's management and her glowing revenue projections, which never seemed to materialize, were beginning to be questioned, in particular by Avie Tevanian, a retired Apple executive who sat on the Theranos board of directors. Holmes responded by threatening him with legal action. Tevanian resigned in 2007, and he warned the other board members that "by not going along 100% 'with the program' they risk[ed] retribution from the Company/Elizabeth."

He was right. Holmes was ruthless about perceived threats and obsessive about company security, and marginalized or fired anyone who failed to deliver or doubted her. Her management was backed up by Theranos chief operating officer and president Ramesh "Sunny" Balwani. Much older than Holmes, he had prospered in the dotcom bubble and seemed to act as her mentor (it later emerged that they were in a secret relationship). To employees, his menacing management style made him Holmes's "enforcer."

Worst of all, Holmes continued to tout untested or nonexistent technology. Her lucrative deals with Safeway and Walgreens depended on her assurance that the Edison could perform over 200 different blood tests, whereas the device could really only do about a dozen. Holmes started a program in 2010 to develop the so-called "miniLab" to perform what she had already promised. She told employees: "The miniLab is the most important thing humanity has ever built." The device ran into serious problems and in fact never worked.

Despite further whistleblowing efforts, Holmes and Balwani lied and maneuvered to keep the truth from investors, business partners, and government agencies. Eminent board members like former U. S. Secretaries of State George Shultz and Henry Kissinger vouched for Holmes, and retired U. S. Marine Corps General James Mattis (now President Trump's Secretary of Defense) praised her "mature" ethical sense. What the board could not verify was the validity of the technology: Holmes had not recruited any directors with the biomedical expertise to oversee and evaluate it. But others were doing just that, as Carreyrou relates in the last part of the book.

In 2014, Carreyrou received a tip from Adam Clapper, a pathologist in Missouri who had helped Carreyrou with an earlier story. Clapper had blogged about his doubts that Theranos could run many tests on just a drop of blood. He heard back from other skeptics and passed on their names to Carreyrou. After multiple tries, Carreyrou struck gold with one Alan Beam, who had just left his job as lab director at Theranos.

After Carreyrou promised him anonymity ("Alan Beam" is a pseudonym), Beam dropped two bombshells. First, the Edisons were highly prone to error and regularly failed quality control tests. Second and more startling, most blood test results reported by Theranos in patient trials did not come from the Edisons but were secretly obtained from standard blood-testing devices. Even these results were untrustworthy: the small Theranos samples had to be diluted to create the bigger liquid volumes required by conventional equipment. This changed the concentrations of the compounds the machines detected, which meant they could not be accurately measured. Beam was worried about the effects of these false results on physicians and patients who depended on them.

Carreyrou knew he had a big story if he could track down supporting evidence. In riveting detail, he recounts how he chased the evidence while Holmes and Balwani worked to derail his efforts. Theranos hired the famously effective and aggressive lawyer David Boies, who tried to stifle Carreyrou and his sources with legal threats and private investigators. Holmes also appealed directly to media magnate Rupert Murdoch, who owned the *WSJ* through its parent company and had invested $125 million in Theranos. Holmes told Murdoch that Carreyrou was using false information that would hurt Theranos, but Murdoch declined to intervene at the *WSJ*.

Carreyrou's front page story in that newspaper, in October 2015, backed up Beam's claims about the Edisons and the secret use of conventional testing. There was an immediate uproar, but Holmes and Balwani fought back, denying the allegations in press releases and personal appearances, and appealing to company loyalty. At one memorable meeting after the story broke, Balwani led hundreds of employees in a defiant chant: "Fuck you, Carreyrou! Fuck you, Carreyrou!"

Problems arose faster than Holmes could deflect them. When Theranos submitted poor clinical data to the FDA, the agency banned the "nanotainer," the tiny tube used for blood samples, from further use. The Centers for Medicare and Medicaid Services, the federal agency that monitors clinical labs, ran inspections that echoed Carreyrou's findings, and banned Theranos from all blood testing. Eventually the company had to invalidate or fix nearly a million blood tests in California and Arizona. In another blow, on March 14, 2018, the Securities and Exchange Commission charged Theranos, Holmes, and Balwani with fraud. Holmes was required to relinquish control over the company and pay a $500,000 fine, and she was barred from holding any office in a public company for 10 years.

Carreyrou tells this intricate story in clear prose and with a momentum worthy of a crime novel. The only flaw, an unavoidable one, is that keeping track of the many characters is not easy— Carreyrou interviewed over 150 people. But he makes sure you know who the moral heroes are of this sad tale.

Two among them are Tyler Shultz, grandson of Theranos board member George Shultz, and Erika Cheung, both recent college grads in biology. While working at Theranos, they noticed severe problems with the blood tests and the company's claims about their accuracy. They got nowhere when they took their concerns to Holmes and Balwani, and to the elder Shultz. After resigning from the company out of conscience, they withstood Theranos's attempts at intimidation and played crucial roles in uncovering what was really going on.

But many with a duty to ask questions did not. A board of directors supposedly exercises "due diligence," which means ensuring that a company's financial picture is sound and that the company's actions do not harm others. The Theranos board seemed little interested in either function, as Carreyrou's story of Tyler Shultz and his grandfather shows. Later on, in a 2017 deposition for an investor's lawsuit against Theranos, the elder Shultz finally did admit his inaction. He testified under oath that, despite escalating allegations, he had believed Holmes's claims about her technology, saying, "That's what I assumed. I didn't probe into it. It didn't occur to me."

Which brings us to the most fascinating part of the story: what power did Elizabeth Holmes have that kept people, experts or not, from simply asking, "Does the technology work?"

Much of the answer comes from Holmes herself. In her appearances and interviews, she comes across as a smart and serious young woman. We learn that her commitment arose partly from her own fear of needles, which of course adds a compelling personal note. Many observers were also gratified that her success came in the notoriously male-oriented Silicon Valley world. Her magnetism was part of what Aswath Damodaran of the NYU Stern School of Business calls the "story" of a business. Theranos's story had the perfect protagonist—an appealing 19-year-old female Stanford dropout passionate about replacing a painful health test with a better, less painful one for the benefit of millions. "With a story this good and a heroine this likable," asks Damodaran in his book *Narrative and Numbers*, "would you want to be the Grinch raising mundane questions about whether the product actually works?"

All this adds up to a combination of charisma and sincere belief in her goals. But Carreyrou has a darker and harsher view: that Holmes's persuasive sincerity was a cover for a master manipulator. Noting her lies about the company finances and technology, her apparent lack of concern for those who might have been harmed by those lies, and her grandiose view of herself as "a modern-day Marie Curie," he concludes his book with this:

> I'll leave it to the psychologists to decide whether Holmes fits the clinical profile [of a sociopath], but [...] her moral compass was badly askew [...] By all accounts, she had a vision that she genuinely believed in [...] But in her all-consuming quest to be the second coming of Steve Jobs [...] she stopped listening to sound advice and began to cut corners. Her ambition was voracious and it brooked no interference. If there was collateral damage on her way to riches and fame, so be it.

I would add one more thought. Holmes did not have the science to judge how hard it would be to realize her dream, then ignored the fact that the dream was failing. Instead, she embraced Silicon Valley culture, which rewards at least the appearance of rapid disruptive change. That may not hurt anyone when the change is peripheral to people's well-being, but it is dangerous when making real products that affect people's health and lives. Facebook's original motto, "Move fast and break things," it seems, is a poor substitute for that old core tenet of medical ethics, "First, do no harm."

*As of this writing (September 2021), Elizabeth Holmes is on trial in federal court, facing charges of several felonies and the possibility of 20 years in prison.

The Bias in the Machine

In January, Robert Williams, an African-American man, was wrongfully arrested due to an inaccurate facial recognition algorithm, a computerized approach that analyzes human faces and identifies them by comparison to database images of known people. He was handcuffed and arrested in front of his family by Detroit police without being told why, then jailed overnight after the police took mugshots, fingerprints, and a DNA sample.

The next day, detectives showed Williams a surveillance video image of an African-American man standing in a store that sells watches. It immediately became clear that he was not Williams. Detailing his arrest in the *Washington Post*, Williams wrote, "The cops looked at each other. I heard one say that 'the computer must have gotten it wrong.'" Williams learned that in investigating a theft from the store, a facial recognition system had tagged his driver's license photo as matching the surveillance image. But the next steps, where investigators first confirm the match, then seek more evidence for an arrest, were poorly done and Williams was brought in. He had to spend 30 hours in jail and post a $1000 bond before he was freed.

What makes the Williams arrest unique is that it received public attention, reports the American Civil Liberties Union [1]. With over 4000 police departments using facial recognition, it is virtually certain that other people have been wrongly implicated in crimes. In 2019, Amara Majeed, a Brown University student, was falsely identified by facial recognition as a suspect in a terrorist bombing in Sri Lanka. The Sri Lankan police retracted the mistake, but not before Majeed received death threats. Even if a person goes free, his or her personal data remain listed among criminal records unless special steps are taken to expunge it.

Recent studies from the National Institute of Standards and Technology and the Massachusetts Institute of Technology [2] have confirmed that computer facial recognition is less accurate at matching African-American faces than Caucasian ones. One reason for the discrepancy is the lack of non-Caucasian faces in datasets from which computer algorithms form a match. The poor representation of people of color from around the world, and their range of facial features and skin shades, creates what researchers have called a "demographic bias" built into the technology.

Facial recognition technology has widespread effects through its association with broad surveillance and massive stores of photographs. In the 1920s, investigators began wiretapping telephones to trace criminal activities. In the 1970s, analog closed-circuit television added remote visual monitoring of people. But digital methods vastly expand the power and scale of surveillance through cameras linked to the Internet and police departments. Ubiquitous in homes, businesses, and public spaces, a billion cameras are projected to be placed in over 50 countries by 2021, one for every eight people on Earth.

To identify suspects, the FBI and police compare images from surveillance cameras and other sources to photo databases. These contain some criminal mugshots, but the bulk of the images comes from non-criminal sources such as passports and state driver's license compilations; that is, the databases mostly expose ordinary, generally innocent citizens to criminal investigation. This approach grew after 9/11, when the U. S. government proposed Total Information Awareness, a global program to collect data about people and identify them by various means, including facial recognition. Georgetown University's Center on Privacy and Technology asserts that half of American adults, 117 million people, appear in databases accessible to police [3]. In 2019, testimony before the U. S. House Oversight Committee revealed that the FBI can scan 640 million photos for facial matching [4].

The FBI and police scan these masses of photos through computer programs that digitize them for identification. An important thread in developing this technology began with the American mathematician and AI pioneer Woodrow Wilson "Woody" Bledsoe. In 1959, he and a colleague invented a machine to recognize alphanumeric characters, then went on to facial recognition.

Their first idea was to analyze a character, say the letter "A," by overlaying it onto a rectangular array of pixels. Each pixel received a binary 1 or 0 depending on whether or not it contained part of the image. The pixels were sampled in adjacent groups called "n-tuples" to account for the spatial relations among them. Further manipulation produced a set of binary digits embodying the letter "A." This process found and stored the bits and a resulting unique score for every character; then an unknown character was identified by comparing its score to the values in memory. The method worked, correctly recognizing up to 95 percent of handwritten and printed numerals.

N-tuples, however, did poorly for the intricacies of a face, whose appearance also varies with illumination, tilt of the head, facial expression, and the subject's age. Bledsoe's team turned instead to human operators who measured characteristic parameters from photographs of faces, such as the distance between the pupils of the eyes or from top to bottom of an ear [5]. In 1967, the researchers showed that a computer using stored facial measurements from several thousand photographs reduced by 99% the number of images a person would have to sift through to match a new photo. Then in 1973, Japanese computer scientist Takeo Kanade automated the entire process with a computer program that extracted eyes, mouth, and so on, from an image of a face without human intervention.

Bledsoe's foundational facial recognition work was funded by the Department of Defense, or according to some evidence, the CIA, either of which would have limited his freedom to publish his results [5]. But early this year, writer Shaun Raviv described in *Wired* what he learned from examining Bledsoe's life and an archive of his work given to the University of Texas after Bledsoe's death in 1995 [6]. The recognition experiments, Raviv reported, began with a database of photos of 400 male Caucasians. In the archive, Raviv saw no references to women or people or color, or images of them in dozens of marked-up photos that must represent Bledsoe's facial measurements.

Since Bledsoe's original research, other techniques have arisen, supported by more powerful computers and bigger databases to develop and test algorithms. Now the introduction of AI methods is bringing about the latest changes; but the bias that comes from the lack of diversity in Bledsoe's formative datasets still appears, and for much the same reason, in these advanced methods.

For years, the U. S. National Institute of Standards and Technology (NIST) has invited producers of facial recognition algorithms to submit them for testing. In 2019, NIST presented its analysis of 189 algorithms from 99 mostly commercial developers [7]. These were checked against federal databases with 18 million images of 8.5 million people for general accuracy and across different demographic groups, in two applications: 1:1 matching, where a face is compared to a stored image for verification, as in confirming the validity of a passport, and 1:n matching, where a face is compared to a whole dataset, typically to find a criminal suspect. For each algorithm, the researchers determined the number of false negatives, where a face that should be matched to one in the database is not, and false positives, where a face is matched to the wrong one.

The data show that facial recognition has improved significantly. The rate of failing to match a submitted face to one in the database dropped from 4% in 2014 to only 0.2% by 2018. Newer algorithms were also less sensitive to the variations in facial appearance that plagued early efforts. The NIST researchers ascribe these gains to an "industrial revolution" in facial recognition, the adoption of deep convolutional neural networks (CNNs).

A neural network is a computing system that can be taught to carry out certain tasks, somewhat like the connected neurons in a biological brain. A CNN mimics human visual perception. In our brains, neurons in specialized regions of the visual cortex register certain general elements in what the eyes see, such as the edges of objects, lines tilted at particular angles, and color. The brain assembles these results into a meaningful whole that allows a person, for example, to quickly recognize a friend even under obscured or varied conditions.

As in the n-tuple method, in a CNN the pixels forming an image are analyzed in spatially adjacent clumps, but succeeding stages provide deeper analysis. Like the regions in the brain, each stage seeks different types of general pictorial elements like those the brain finds, rather than seeking the eyes, nose, and so on. The mathematically manipulated results are passed on and augmented through the stages, finally producing an integrated representation of a face. Crucially, this is achieved by first exposing the CNN to a large dataset of varied facial images. This "trains" the system to develop a comprehensive approach to analyzing faces.

Within NIST's testing, CNN-based algorithms performed best; but overall, the algorithms differed in how well they identified people of different races, sexes, and ages. These results echo earlier studies of 1:1 matching and are the first to explore demographic effects in 1:n matching. Errors in each application yield different undesirable outcomes. A false positive in a 1:1 search can allow unauthorized access; a false positive in a 1:n search for a criminal suspect puts the subject at risk for unwarranted accusations.

In 1:n matching, the NIST data show that the most accurate algorithms are also the most reliable across demographic groups. Less proficient ones give higher rates of false positives for African-American females compared to African-American males, and to white males and females, in an FBI database of 1.6 million mugshots. For 1:1 matching, some algorithms falsely matched African-American and Asian faces 10 to 100 times more often than Caucasian ones. Notably, however, some algorithms from Asian countries gave fewer false positives for Asians than for Caucasians. This, the report notes, shows that the degree of diversity in a training dataset may strongly affect the demographic performance of a CNN.

Other research has more fully explored how lack of diversity affects the training of a neural network. In 2012, B. F. Klare and A. K. Jain at the Michigan State University and colleagues tested 1:1 facial matching against police mugshots [8]. Different types of algorithms they examined were all less accurate for African-American faces than white or Hispanic ones. One algorithm studied was a neural network defined by its training dataset. The researchers found that the resulting fits to African-Americans improved when this dataset was limited to African-American faces, and in a nod to diversity, also improved when the training dataset had equal numbers of African-American, Hispanic, and white faces.

This suggests how to make biased training databases more equitable. In one recent demonstration, researchers at the biometrics company Onfido made a demographically unbalanced dataset less biased [9]. Its facial images came from different continents in varying proportions, such as 0.5% from Africa compared to 61% from Europe. This yielded a false positive rate 63 times higher for African faces than for European ones. But when the researchers used statistical methods to train with more African faces than their small numbers alone would provide, the discrepancy was reduced to a factor of 2.5, a sign of future possibilities.

But according to biometrician Patrick Grother, lead scientist for the NIST report, serious police action should require more than just a match from an algorithm. He explained that an algorithm actually returns a list of likely candidates. In the ideal next step, an investigator seeking suspects must confirm that there is a good match within this list. Only then would the detective seek other evidence like eyewitnesses or forensic data to justify arresting and charging the subject. The fact that a "no match" from a human investigator can overturn a wrong machine identification should be reassuring, but that came too late to save Williams from false arrest and its repercussions.

Andrew Guthrie Ferguson is a professor at American University Washington College of Law who studies technology and civil rights. Responding to my query, he wrote that "facial recognition should not be used to deprive people of liberty." It is "too dangerous a tool to be used in an unregulated way. Williams' case is a signal to stop the *ad hoc* adoption of facial recognition before an injustice occurs that cannot be undone."

Repairing the flaws in facial recognition technology will not be easy within a complex landscape that includes dozens of producers of the software with varying levels of bias, and thousands of law enforcement agencies that can choose any of these algorithms. Maybe only a federal effort to establish standards and regulate compliance to them would be necessary before we no longer have a Robert Williams, a member of any minority group, or any citizen unjustly experience a night in jail or worse.

References

1. Garvie, C. The untold number of people implicated in crimes they didn't commit because of face recognition. aclu.org/news (2020).

2. Buolamwini, J. and Gebru, T. Gender shades: Intersectional accuracy disparities in commercial gender classification. *Proceedings of Machine Learning Research* **81**, 1–15 (2008).

3. Garvie, C., Bedoya, A., and Frankle, J. The perpetual line-up. Georgetown Law Center on Privacy & Technology (2016).

4. Melton, M. Government Watchdog Questions FBI on Its 640-Million-Photo Facial Recognition Database. *Forbes* (2019).

5. Boyer, R. S. (Ed.) *Automated Reasoning: Essay in Honor of Woody Bledsoe.* Kluwer Academic Publishers, Dordrecht, Netherlands (1991).

6. Raviv, S. The Secret History of Facial Recognition. *Wired* (2020).

7. Grother, P., Ngan, M., and Hanaoka, K. Face recognition vendor test. National Institute of Standards and Technology (2018).

8. Klare, B. F., Burge, M. J., Klontz, J. C., Vorder Bruegge, R. W., and Jain, A. K. Face recognition performance: Role of demographic information. *IEEE Transactions on Information Forensics and Security* **7**, 1789–1801 (2012).

9. Bruveris, M., et al., Reducing geographic performance differentials for face recognition. *arXiv* (2020). Retrieved from DOI: 2002.12093.

The War Science Waged

As I watched the breathtaking TV images of our missile attacks in the Gulf War, I was as amazed as everyone else; but I felt another less common emotion as well, the lifting of guilt. The guilt had been there because my scientific research has helped to weave the intricate technological web that supports modern warfare. For me, this research raised a moral question, whose answer the Gulf War has provided.

Unlike those scientists who struggled with the rightness of building the atomic and hydrogen bombs, I'm not a nuclear physicist or bomb designer, but a solid-state physicist who studies semiconductors. You'll find my small research laboratory on a serene university campus, not at Los Alamos or other huge government installation. In that lab there are only optical instruments and a few researchers, tools and people to wrest nuggets of understanding from shiny chunks of semiconductors.

What do these bits of material have to do with the Gulf War? A great deal. War-making today depends on electronics and photonics, the sciences that manipulate electrons and light, as much as on tanks and missiles. These technologies are critical to detect and track targets in real time, to guide weaponry to them, jam enemy radar, photograph bombed targets for damage assessment, and plan elaborate logistical support for troops and air sorties.

Some of these devices are like items we all use daily. But military electronics and photonics also need devices using exotic new materials not yet established in our commercial economy. Cases in point are the semiconducting compounds indium antimonide and mercury cadmium telluride, which respond to infrared light. Infrared sensors made of these or other materials can find and track targets by their invisible emissions, to eventually steer a missile to impact or guide infantry through darkness.

In scientific areas of military importance, the research funding from the Defense Department can far exceed civilian resources. And so, many academic scientists have focused efforts on the properties of exotic semiconductors, lured by the chance to get a piece of The Pentagon pie. This does not mean that the resulting research has no scientific value. On the contrary, it addresses important questions and may open up applications of real value to peaceful society, as has happened many times before.

Nevertheless, scientists like me on tree-lined campuses, with no connection to H-bombs and other nightmare weapons, have contributed directly to the military potential of the United States. I've had my doubts about this. Remembering the courage of some in denouncing military use of science, and remembering too my own liberal upbringing, I haven't liked knowing that some of this work might allow better targeting for our weapons. And now the Gulf War has recast my thinking.

If technology is itself neutral, or equally capable of good and bad consequences, its morality is determined by its political use. As much as I can tell—and my sources are no better than any other citizen's—America's strategy was to put a sharp technological spear point on war's blunt bludgeon. The means are not perfect; errors happen. Civilian sites may be misidentified as military targets, or targets may be hidden in civilian surroundings; even when military targets are attacked, unarmed people may die. But to me, our stated goal of minimizing innocent casualties, along with minimizing casualties for our own soldiers, has moral weight. And it is the technical means to select targets that make the moral decision possible, and therefore meaningful. We have come some distance from the blanket fire-bombing of Dresden in World War II.

I admit to a worm of doubt. When I watch Gens. Colin L. Powell and H. Norman Schwarzkopf, both looking able, confident, and honorable, I see in overlay the face of Gen. William Westmoreland. He too looked every inch the capable and decorated warrior, as he assured us that we were winning in Vietnam. And with this memory, I wonder how many civilian casualties we caused in Iraq. I also understand that technology alone did not win the war. But if laboratory research has produced exotic chips that save lives on our side and on theirs, I have contributed to good technology.

From *Terminator* to *Black Mirror*: Algorithmic Warfare's Perils

Reviews of the films *Sleep Dealer, Good Kill, Eye in the Sky, Slaughterbots*, and of the TV segment *Black Mirror: Metalhead.*

Since James Cameron's *Terminator* android (Arnold Schwarzenegger) tracked and murdered human targets in 1984, we've seen a real-life, accelerated evolution in artificial intelligence (AI) that could be equally threatening. Technological advances in remote or autonomous surveillance, identification, and delivery of lethal force inspire considerable debate over their uses by law enforcement and the military, especially in drone warfare.

Much of the debate is about facial recognition AI, which identifies a subject by comparing their face to enormous databases of known faces. Police use it to find criminal suspects by matching their faces to surveillance videos, but in today's search for social and racial justice, the algorithms are severely criticized. Their accuracy is not subject to any standards, and they are more error prone for non-Caucasian faces than Caucasian ones. These flaws became real and prominent this year, when an incorrect algorithmic facial identification led Detroit police to falsely arrest an African-American man. He was released only after spending 30 hours in jail and posting a bond.

Correctly identifying a person is essential for equitable policing, and that extends to the even more potentially destructive use of drone warfare, in which the U. S. military finds and kills enemy combatants with armed semiautonomous drones. These are controlled by remote operators thousands of miles distant, subject to higher authority. Reports indicate that the military may soon add facial recognition to its drones to reduce human error, but given the lacks in the technology, this may only prove a complication.

With or without recognition technology, however, the human links in the decision chain are meant to provide oversight and final approval for lethal drone attacks. This has not prevented U. S. drones from killing or injuring at least 200 civilians in Iraq and other war zones in 2019. Recognizing this issue, several filmmakers have recently addressed our current qualms with AI and facial recognition. These films portray people and technology interacting in the use of killer drones.

Director Alex Rivera's near-future film *Sleep Dealer* (2008) follows Memo (Luis Fernando Peña), whose father, a Mexican farmer, is killed by a drone strike for protesting a corporation's dam, built to sell water for profit. After his father's death, Memo finds work in Tijuana, "jacking in" to a neural network to remotely build a skyscraper in the U. S. When he meets Rudy, the drone operator who killed his father, they come to an understanding and unite to smash open the dam and free the water for the community. Ultimately though, they cannot destroy the corporation itself. In this intervention, Rudy rises above "just following orders" to act on what he feels is justice.

In Andrew Niccol's *Good Kill* (2014), another drone operator faces up to his conscience. U. S. Air Force Major Thomas Egan (Ethan Hawke) proficiently kills terrorists in Afghanistan from a drone base in Nevada. But he comes to feel guilty, especially under a CIA manager who finds civilian casualties acceptable. Egan drinks heavily and his marriage suffers. To redeem himself, he simulates a drone malfunction to let civilians on the ground escape, then uses the drone to kill a known rapist who surveillance shows is once again approaching a former victim. The film ends on an unresolved note, showing Egan leaving his post for an unknown fate. Like Rudy, Egan counters orders to use the drone to fulfill his personal choices, but letting his emotions drive him to carry out a summary execution is not a moral response.

Eye in the Sky (2015), a thriller directed by Gavin Hood, shows how dependence on drone technology and AI can affect human judgment. British Army Colonel Katherine Powell (Helen Mirren) plans to observe and capture terrorists in Nairobi, Kenya, using drones controlled from the U. S. When a facial recognition algorithm identifies suicide bomber terrorists who could kill civilians, she changes the drone's goal to "kill." Besides her complete trust in the AI identification, this is legally and morally questionable: The U. K. and U. S. are not at war with Kenya, and a drone strike could harm a nearby young girl. Zealous to gain government approval for lethal action, Powell deceitfully reports the odds of killing the girl as below 50%. After authorization by the U. S. Secretary of State, the drone operator fires missiles that kill the terrorists but introduce moral ambiguity by also killing the girl.

Figure 2   In the thriller *Eye in the Sky* (2015), British Army Colonel Katherine Powell (Helen Mirren) underplays the collateral damage when she calls for authorization to kill terrorist bombers by a remote drone strike. The drone attack kills an innocent young girl along with the terrorists.

Overall, these three films show that human choices can sometimes properly modify or override technological decisions. In two dystopian works that anticipate the future of machine killing, the human element is removed, with bleak results.

The short film *Slaughterbots* (2017) begins with a man on stage before a live audience. He looks like a tech company executive, but he is selling a weapon: a palm-sized, autonomous drone armed with facial recognition AI and an explosive charge to efficiently find and kill a victim. With fictional news clips and interviews of distraught people, the film imagines how chaos could result from unstoppable swarms of the drones. *Slaughterbots* is effectively a call to action from the Future of Life Institute, which aims to reduce threats to humanity from AI. The film's dramatic tone highlights the dangers of autonomous weapons in large quantities.

How close are we to facing the horrors of the autonomous drones in *Slaughterbots?* Not very close. The armed U. S. military Reaper drones are big, long-range units, with a wingspan of up to 79 feet. *Slaughterbots* has inspired discussions that explain how far off we are from creating the necessary AI and cramming it into tiny drones along with an explosive charge and extended flight capability.

Made the same year as *Slaughterbots*, the *Black Mirror* episode *Metalhead* (2017) imagines the dangers of autonomous technology differently. In an undefined future, Bella (Maxine Peak) and two male companions break into a huge warehouse, searching for something until their movements awaken a watchdog-like robot with an unnervingly featureless head. It shoots small devices into the intruders to tag them, brutally kills the men, then follows Bella as she flees through open country and into a vacant house. The robot dog is fast, strong, and smart, but Bella uses her human skills and fortitude to painfully cut the embedded tags from her flesh, then disables the dog with shotgun blasts.

Still, she and humanity do not really win. In its last gasp, the dog shoots more tracking devices into Bella. The final scene shows her with knife in hand, hopeless and ready to slit her throat as more robot dogs converge on the house. The episode hints that the dogs have already hunted down most living things, leading a viewer to speculate that they are killer robots left over from a previous war.

As an entry to the world of *Metalhead,* we already have ground-based robots with substantial physical abilities. The dog in *Metalhead* resembles actual robots created by the company Boston Dynamics, although battery capacity limits their potential. However, to chase Bella, the dog had to perform high-level cognition such as navigating complex environments and spontaneously deciding to use a kitchen knife as a weapon. Today, AI performs some tasks better than people, but generalized AI does not yet match the cognitive abilities of a real dog or person.

The errors and biases in facial recognition are good reasons to be wary of it and other AI being applied in warfare. But whatever the intelligence of future self-guided weapons, they would be relentless like the Terminator which, as described in the 1984 film, "can't be bargained with. It can't be reasoned with. It doesn't feel pity, or remorse, or fear!" For all their flaws, people in the decision chain feel those emotions, and therefore remain necessary to keep warfare by algorithm from becoming an inhuman nightmare.

Radiations: What Came Out of the Atomic Bomb

Review of *Atomic Doctors Conscience and Complicity at the Dawn of the Nuclear Age*, James L. Nolan. Jr. (Harvard University Press, 2020).

Direct memories of the atomic bomb are fading 75 years after World War II ended, but many personal links remain. I was a young student physicist at the Los Alamos lab in 1960, where I encountered some of the scientists who had created the bomb. Like them, I breathed the clear air at 7300 feet, which seemed to sharpen perception and perhaps honed the intensive effort to build the bomb. In 2002, I was allowed to visit the Trinity site at Alamogordo, New Mexico, where the first atomic bomb was tested. Again like those original scientists, I saw a wasteland where the heat of the explosion had fused sand into glass. This foreshadowed the desolation at Hiroshima and Nagasaki after the bombs were dropped, ending the war and changing the world.

My limited connections give me a feel for the place and the people, but deeper links give deeper insights. In *Atomic Doctors*, James L. Nolan Jr., a sociologist of technology at Williams College, draws on personal materials handed down from his grandfather, physician and U. S. Army Captain James F. Nolan. From this unique viewpoint, the book illuminates how Dr. Nolan at Los Alamos and two physician colleagues, Louis Hempelmann and Stafford Warren, dealt with the frightening human effects of nuclear radiation from the bomb. Combining an effective analysis of their efforts with a compelling telling of Dr. Nolan's own story, the book enlarges America's atomic bomb experience as a case study of truly disruptive technology in war and society.

The physicians came to the bomb project with different allegiances. Physicist Robert Oppenheimer, the scientific chief at Los Alamos, recruited Nolan and Hempelmann to build a general medical facility and monitor radioactive materials. General Leslie Groves, who ran the entire Manhattan Project including Los Alamos, brought in Warren as the Project's chief medical officer. With Nolan, he was inducted into the army. Nolan did not enjoy his military role, but Warren relished his rank of colonel and was directly responsible to Groves. Hempelmann, however, remained a civilian and oversaw radiation safety in lab areas.

These differences reflected the mismatched cultures at Los Alamos. The military was dedicated to producing a decisive weapon. The scientists were caught up in exploring the science behind the bomb and in the technical challenge of making it work. Used to open academic research, they bristled at military discipline and secrecy. Groves returned the favor, calling the scientists "the largest collection of crackpots ever seen" (although he and Oppenheimer miraculously managed to work well together).

Neither culture fully accepted the physicians. World-class scientists made their own medical diagnoses and argued with the doctors; the military simply wanted to "explode the damn thing" and did not worry about radiation. They saw the bomb narrowly as producing heat and blast like conventional explosives, only more so; its destructive power at Nagasaki was described as that of 20,000 tons of TNT. But the physicians had their own ethical and professional obligations to warn of the harm that radiation would bring.

Before the Trinity test on July 16, 1945, of the Gadget, a plutonium-based bomb, Hempelmann and Nolan were unable to raise awareness about its radiation dangers until two lab scientists, Joseph Hirschfelder and John Magee, supported them. The group calculated the expected radiation levels and wrote a report about the need to protect lab personnel and towns near the Trinity site from radioactive dust.

On June 17, 1945, Nolan hand carried the report to Groves at Oak Ridge, Tennessee. Groves dismissed it, fearing that leaks about dangers to civilians would compromise secrecy, and refused to wait for weather that would limit the spread of radioactive debris. He was under heavy pressure to deliver a successful result to President Harry Truman by July 17, when Truman would meet Winston Churchill and Joseph Stalin at Potsdam, Germany. However, Warren eventually persuaded Groves to allow some limited safety efforts at Trinity.

Trinity was a military success. The bomb was more powerful than expected, but also produced significant radiation and fallout as predicted; for instance, 20 roentgens were measured in one canyon 20 miles north of ground zero, far above what was then thought to be a safe level. Fifty miles north, 12 girls at a summer camp played with warm "white snow" drifting down from the sky. Reportedly 10 of them died of cancer before age 30. Despite these and other signs that radiation harmed local people, the U. S. government has never compensated any of these so-called Trinity downwinders.

Instead of participating in the follow-up analysis, Dr. Nolan was aboard the cruiser *USS Indianapolis* with a heavy lead canister containing most of the uranium-235 for Little Boy, the bomb that destroyed Hiroshima. He had been chosen to monitor radiation safety as he and another officer escorted the heart of the bomb to Tinian Island in the Pacific Ocean, home base of the *Enola Gay* B-29 bomber that dropped the bomb on August 6, 1945.

Here the author's access to family records humanizes the story with a few comic and sadly paradoxical moments. To preserve secrecy, Dr. Nolan was given a cover identity as an army artillery officer. The disguise was hardly well thought out. Nolan mistakenly wore his new insignia upside down and had to nervously dodge technical questions about artillery from naval officers. Besides, as a reluctant military officer who didn't even want to handle his sidearm, it was ironic that he played a direct role in the first use of the atomic bomb in war. After writing that a friend asked Dr. Nolan "Does this make you a hero or a villain?" Nolan's grandson adds, "I suspect my grandfather quietly pondered this provocative question for years after the war."

After Hiroshima, and the Nagasaki bombing on August 9, Nolan and others suggested that their aftermaths should be studied on the ground in Japan. An investigative commission was assembled, and a Los Alamos contingent, including Nolan and Warren, was its first component to enter Japan. Groves wanted the team to counteract leaks about radiation sickness at Hiroshima, such as a covert news report of deaths due to an "atomic plague." Radiation was like biological and chemical attacks, which were illegal as inhumane forms of warfare, but more mysterious and insidious. The human effects of neutrons and gamma rays from the explosion and of residual radiation like that at Trinity were poorly understood. The military wanted to claim instead that deaths and injuries were due to blast and heat.

On September 9, Nolan and Warren with their group walked through Hiroshima, levelled so that at nearly a mile from the epicenter there was simply nothing to see. They visited improvised hospitals where a hundred Japanese were dying daily, and one physician with the group described "thousands of helpless, suffering bodies stretched out on the ground; the thousands of swollen charred faces; the ulcerated backs; the suppurating arms." This was not what Groves wanted to hear. Nevertheless, Nolan and Warren found evidence of direct harm from gamma rays, and of residual radioactivity, at both bombed cities. But comments that Warren made through military channels minimized or contradicted the data, as in his claim that "there never was a dangerous amount of radiation anywhere in the city of Nagasaki or its environs."

After returning home in Fall 1945, Nolan left the military and became the civilian head of the Los Alamos Health Group. Meanwhile Warren delivered a report of his and Nolan's findings in Japan to Groves, which *Atomic Doctors* describes as written to favor the General's desired narrative. Even so, when Groves testified before the U. S. Senate in November 1945, he contradicted some of the report's medical conclusions but also appealed to medical authority depending on which helped his case.

He asserted, for example, that the number of Japanese deaths due to direct radiation was small, whereas the report noted a 33% rate of radiation effects among 4000 patients, with half the irradiated people dying. But Groves also claimed that according to the doctors, radiation provides a "very pleasant way to die." He knew, however, of the miserable lingering death of Los Alamos physicist Harry Daghlian after a radiation accident in August 1945, with symptoms like those seen in Japan. One Senator noted that Groves seemingly protested too much about the lack of radiation effects, perhaps because he was defensive about inflicting such harm.

Meanwhile studies of the human effects of radiation had produced another fraught issue. After the accidental plutonium poisoning of a Los Alamos scientist in 1944, Hempelmann suggested a program to determine how long plutonium lingered within the body. Oppenheimer approved this even to the point of human experimentation, although the project was kept secret. The first suitable subject was identified as a "colored male" named Ebb Cade. A cement worker at the Oak Ridge lab, he was recuperating from an auto accident that broke an arm and a leg. In a disturbing reminder of the notorious Tuskegee syphilis experiments on African-American men, on April 10, 1945, Cade was injected with 4.7 micrograms of plutonium without his knowledge or consent, and various bodily samples were sent to Los Alamos for analysis.

Seventeen other patients were likewise injected at several facilities, but none were clearly told what was being done to them and only one gave consent. Finally, in 1947, the new civilian Atomic Energy Commission introduced restrictions on future human experiments. The program, however, remained secret until it was revealed in the 1990s.

The plutonium experiments, writes the author, were the most extreme example of how doctors were co-opted and made complicit with Trinity, radiation accidents, and radiation effects in Japan. The doctors gave warnings that the military ignored or dismissed. When the doctors gathered data that confirmed the warnings, the military response was to misrepresent any results that could not be kept secret. The doctors accepted these acts from fear of litigation.

What I take away from this summary is that the first atomic bombs produced more than radiation. The responses to its deadly effects definitively established that when objective analysis threatens political goals, doctors and scientists can do little against the power of government to silence them and shape public opinion. That lesson still applies as our government uses these tools against unwelcome scientific warnings about the coronavirus pandemic and climate change.

Near its end, the book puts the atomic bomb experience into the larger context of how to weigh technology's potential for good or harm by presenting different ways to consider the question. One approach is instrumentalism, which holds that technology is value free but that its outcomes depend on the values that govern its use. Dr. Nolan's own life, writes his grandson, embodied this view. After leaving Los Alamos, Nolan became a professor of gynecology and obstetrics at the University of Southern California and director of a cancer center. This reflected his specialty in treating cancer with radiation, an application that drew on different values than did the atomic bomb project, and saved lives.

These different ethical frameworks played out inside Nolan as well. Despite his role in warning against the harm from radiation, and the moral questions it raised, he never changed his opinion that the atomic bombings were necessary and saved American lives.

The tension between these positions is expressed in his grandson's comment about Nolan pondering whether he was a hero or a villain. Family members saw other signs of internal struggles. Nolan's daughter Lynne knew that what he had witnessed in Japan troubled him, saying "It's got to leave a mark...he would go through funks...he walked in there right after the bomb went off and saw people dying. Not only the dead, but the dying."

Dr. Nolan could not fully resolve his own ethical contradictions. Today, even when nuclear war would go far beyond what was imaginable in 1945, neither can our whole society. *Atomic Doctors* does not tell us how to do this, but powerfully shows that it must be done.

The Physics of Blood Spatter

Joe Bryan, once a popular and respected high-school principal in a small Texas town, has been in prison for over 30 years. He is serving a 99-year sentence for the shooting and murder of his wife in 1985.* The evidence incriminating him involved spots of the victim's blood found on a hand-held torch.** A witness, who was rated as expert in the forensic technique of blood pattern analysis (BPA), interpreted these spots as placing Bryan near his wife when she was shot—a testimony that was at the forefront of Bryan's conviction. It overrode countervailing evidence that he was in fact at a conference 120 miles away—an alibi that made it nearly impossible for him to have shot his wife, as he would have had to leave the event, travel home, commit murder, and return to the conference within a specific timeframe. Bryan maintains his innocence to this day.

Evidence like this, based on the physical behavior of the blood generated at a crime scene, has roots in late 19th-century Europe. It became prominent in the U. S. during the famous Sam Sheppard murder trial in 1955, and has played an important role in other murder trials since—including those of football player and actor O. J. Simpson (1994–1995, verdict of not guilty) and music producer Phil Spector (2007–2009, retrial verdict of guilty).

Police investigators use BPA to work backward from blood traces at a crime scene, allowing them to reconstruct the locations and actions of the people and weapons involved. The traces include drips, smears, and spatters, which are created when drops of blood radiate from the impact of a bullet or blunt instrument until they encounter a surface and stain it. But according to a startling 2009 report from the U. S. National Academy of Sciences (NAS) that still resonates today, BPA lacks scientific rigor and valid accreditation for its practitioners. This is a serious concern because BPA results have convicted people later shown to be innocent, as many believe Bryan to be, and because lack of confidence in BPA analysis may allow the guilty to go free. As a result, it has become essential to re-assess the physics behind BPA.

Although the U. S. leads the world in gun ownership—there are 120 guns per 100 people, and 64% of U. S. homicides are gun related—other countries have many shootings too. For instance,

30% of homicides in Canada involve guns, while a dozen nations, including Brazil, exceed the U. S. in their rate of gun deaths per 100,000 people. Establishing the scientific validity of BPA could, therefore, have an international impact on dealing with the world's 250,000 annual gun-related deaths by helping to categorize them as homicides or suicides—and, in the former case, potentially bringing the perpetrators to justice.

Bloody behavior

In terms of physics, BPA reconstruction is a complicated problem in fluid mechanics that involves tracing the behavior of blood, under various forces and ambient conditions. The challenge is made more difficult because blood is a complex fluid containing both liquid (the plasma) and solid (the blood cells) components. Furthermore, the properties of blood, such as its pH or the number of red blood cells, vary from person to person.

But this work is more than just an academic exercise. It can also have real effects, according to Alicia Carriquiry at Iowa State University in the U. S. As a statistician and director of the Center for Statistics and Applications in Forensic Evidence (CSAFE), which is funded by the U. S. National Institute of Standards and Technology (NIST), Carriquiry has a broad view of forensic science. "BPA is one of those areas in which science has a lot to say," she says. "As opposed to other forensic disciplines, in BPA we actually have physical and fluid dynamical models that can help answer questions such as those having to do with trajectory, point of origin and similar."

Fundamental science has, however, not been well applied to BPA, according to the 2009 NAS report, which was entitled *Strengthening Forensic Science in the United States.* Co-chaired by a distinguished U. S. federal judge and an academic statistician, it included contributors from relevant scientific disciplines, including physics. In general, except for DNA analysis, the report found deficiencies in virtually every forensic technique, including the analysis of hair, fibers, fingerprints, and bite marks. "The interpretation of forensic evidence is not always based on scientific studies to determine its validity," it stated. "This is a serious problem." A separate report in 2016 from the U. S. President's Council of Advisors on Science and

Technology—written by the director of the Office of Science and Technology Policy together with a panel of scientists—echoed this critique.

For BPA in particular, the NAS report noted the complexities of fluid dynamics and indicated that BPA analysts should understand the physics involved. But with no strict educational requirements for certification as a BPA expert—they're trained only to follow packaged procedures—the report concluded that "The opinions of bloodstain pattern analysts are more subjective than scientific... The uncertainties associated with [BPA] are enormous." In 2018, the Texas Forensic Science Commission reached similar conclusions about the Bryan case, calling the interpretation of the BPA evidence "inaccurate" and "scientifically unsupportable."

Still, properly used, BPA can give valuable clues toward understanding the circumstances of a shooting. For example, drops of blood that strike the floor at an angle will create a set of elliptical stains, whose width-to-length ratio gives the impact angle. BPA analysts are trained to draw straight-line trajectories that follow the long dimension of each ellipse at that impact angle. These paths converge, providing a position from which the blood originated. While this correctly gives the projection onto the floor of the location of a gunshot wound, the straight-line procedure overestimates the height of the wound, since the true paths under gravity are parabolas modified by aerodynamic drag. The error is typically large enough to wrongly place a victim as standing rather than sitting.

This is one of the established BPA methods that deeper physical analysis can improve. In 2011, physicists Christopher Varney and Fred Gittes at Washington State University put the projectile-motion equations, including gravity and drag, into a form that uses all the data inherent in a set of spatter bloodstains [1]. They found that a plot of the impact angles for the stains *versus* the inverse of their horizontal distances from the vertical axis of impact gives a valid result for the height, provided that the launch angles for the drops are not too widely distributed. In a test that spattered a viscous blood substitute, the researchers used this approach to calculate the actual launch height of 88 cm to within 8%. For comparison, the linear trajectories overestimated the launch height by 100%.

In 2015, Nick Laan, of the University of Amsterdam and the Netherlands Forensics Institute, and colleagues instead used the fluid qualities of blood to find the height of a gunshot wound [2]. Earlier work had derived an equation relating the impact velocity of a liquid drop of blood to its volume and impact angle, and to the width of the dried stain it produced as determined by the known capillary and viscous behavior of blood. To apply this method, the researchers created spatter patterns of human blood under controlled conditions. For each of 40 separate blood stains, they determined its width and impact angle, and using a commercial 3D surface scanner, measured the volume of the stain, from which they found the volume of the original drop. These parameters yielded the impact velocity, giving enough information to solve the equations of motion under gravity with aerodynamic drag. The results for the height where each drop originated had an average value of 58.5 cm, only 8% below the true height of 63.7 cm. Meanwhile, the straight-line method gave 91.1 cm—a much larger 42% error.

A bullet's journey

These two papers and others analyze the behavior of blood drops after they have been formed, to enhance standard BPA. But mechanical engineers and fluid dynamicists Alexander Yarin and Patrick Comiskey from the University of Illinois in Chicago, together with Daniel Attinger at Iowa State, have gone further. They have modeled the entire process from the bullet entering the body to the final blood stain pattern.

Since 2016 these researchers have developed fluid dynamical theories for gunshot back spatter and forward spatter, where blood drops travel, respectively, against and with the direction of the bullet and display different characteristics. The back-spatter analysis—carried out for both regular [3] and blunt-nosed [4] bullets—is based on the well-known Rayleigh–Taylor instability. In this effect, acceleration perpendicular to the interface between two fluids of different densities—here, blood and air—creates growing turbulence and mixing between the fluids. (One remarkable example of the instability is the spectacular filaments seen in the expanding Crab Nebula, where the two fluids comprise material ejected by the Crab's initial supernova explosion, and a plasma of relativistic charged particles powered by the Crab's central pulsar.) Meanwhile, forward spatter, caused when the bullet exits a body after multiple disruptive encounters with blood and tissue, was treated differently. Its analysis used percolation theory, which describes available paths through randomly arranged clusters.

For both kinds of spatter, the researchers calculated the numbers, sizes, and dynamical properties of the drops of blood generated by a bullet; then determined their trajectories under gravity and aerodynamic drag. Finally, the team found the number of resulting stains, their areas, impact angles, and distribution with respect to distance.

These calculated results for the blood-spatter distributions agree reasonably well with data obtained by shooting bullets into sponges or plastic foam soaked with swine blood for back spatter, and through a blood-filled reservoir for forward spatter. Although the researchers note that more experiments are needed, their results are significant steps toward a real, physics-based theory of spattering. Their work also points to new directions for study, such as how air is carried along with drops of blood in flight, which influences their trajectories, and the impact of temperature on blood viscosity.

The theoretical results so far show the value of the fluid dynamics approach but also that its complexity can add uncertainties to the analysis, for example through the variable properties of blood and blood stains. Besides temperature, the viscosity of blood also depends on the percentage of red blood cells, which varies by individual and could affect approaches like that used by Laan and colleagues. What's more, the properties of the surface that a blood drop strikes may modify how it spreads and, therefore, affect the stain it leaves. Confounding elements like these should be taken into account for fully valid BPA that carries weight in court and may limit claims about what BPA can definitively show. Certainly, there is much left to do.

From lab to crime scene

As the science of BPA progresses, a parallel challenge is to convert its results into new, practical, and transparent procedures for murder investigations and courtroom presentations. But BPA practitioners have not entirely welcomed these changes, which threaten to upset established field procedures, a reaction also found elsewhere in the forensics establishment. Nevertheless, says Carriquiry, for many topics in forensic science "we have managed to make important inroads and established some meaningful partnerships with forensic practitioners who see our work as a means to make theirs more objective and 'scientific.'" These partners include the Houston Forensic Science Center and the Los Angeles Police Department.

Now, with support from the CSAFE and other U. S. federal agencies, researchers are working specifically to strengthen connections between the BPA and fluid dynamics communities, and to provide practitioners with useful results. For example, Attinger and co-authors have written a tutorial paper in which they discuss the forces at play in fluid dynamics and how they determine the behavior of blood drops at a crime scene [5]. Attinger has also published charts based on fluid dynamics that make it simple for investigators in the field to estimate the maximum distance that a blood drop has traveled [6].

In another effort, Attinger's team has published back-spatter patterns of human blood produced in the lab by gunshot, with rigorous control of the firearms and ammunition used and the physical arrangement [7]; and a second set of blood patterns produced by blunt instruments [8]. These provide high-resolution images of blood stains generated under varied conditions, for training and research use. In one project, Hal Stern at the University of California–Irvine, a statistician and co-director of CSAFE, is examining the images for distinctive features that practitioners could use to distinguish among possible sources for observed spatters. In other outreach, BPA researchers also present talks and training sessions at professional societies.

Unfortunately, widespread adoption of more rigorous BPA practice and training will not come quickly, or automatically erase past deficiencies that produced unreliable evidence and false accusations. Nor is it likely that the legal standards for acceptance of BPA evidence will change soon enough to affect Joe Bryan's upcoming appeal for a new trial. That request was denied in 2018, but his lawyers are now preparing a last-ditch effort before the Texas

Court of Criminal Appeals. However, Bryan is nearly 80. Even if a new trial is approved, it may not come in time to do him any good.*

Whatever that outcome, the extensive coverage of the Bryan case along with the NAS report and other evaluations of BPA have uncovered its problems and motivated progress toward a better physics of blood patterns. This may at least ensure that future blood evidence will be more effective in identifying the true perpetrators without unjustly condemning people who are innocent.

———————

*After further consideration of the blood-spatter errors found in his trial, on March 21, 2020, Joe Bryan was released from prison on parole, 33 years after his conviction for murder. He was 79 years old.

**"Torch" in U. K. usage is "flashlight" in U. S. usage.

References

1. C. R. Varney and F. Gittes, *Am. J. Phys.* **79**, 8 (2011), 838–842.

2. N. Laan et al., *Sci. Rep.* **5**, 11461 (2015).

3. P. M. Comiskey et al., *Phys. Rev. Fluids* **1**, 043201 (2016).

4. P. M. Comiskey, A. L. Yarin, and D. Attinger, *Phys. Rev. Fluids* **2**, 073906 (2017).

5. D. Attinger et al., *Forensic. Sci. Int.* **231** (2013), 375–396.

6. D. Attinger, *Forensic Sci. Int.* **298** (2019), 97–105.

7. D. Attinger et al., *Data in Brief* **22** (2019), 269–278.

8. D. Attinger et al., *Data in Brief* **18**, 648–654.

From the Lab to the Courtroom

Not many physicists carry a gun to defend themselves against attackers provoked by their research, but that's exactly what Wilmer Souder once felt the need to do. Since 1911 he worked at the U. S. National Bureau of Standards (NBS) in Washington, DC (today it is the National Institute of Standards and Technology (NIST)), eventually developing forensic techniques that convicted criminals. Souder was not the only forensic physicist in that era. John H. Fisher, another ex-NBS physicist, invented a device essential for forensic firearms identification. Both of their contributions were important in major criminal trials and made a sizable impact on the justice system.

Fisher worked at the independent Bureau of Forensic Ballistics, established in 1925, where he invented the helixometer to peer inside the barrel of a firearm without sawing it in half lengthwise. His patent shows the device's optical arrangement and graduated angular scale that allowed an investigator to examine defects in the barrel and find the pitch of its rifling—the internal spiral groove that imparts a stabilizing spin to a bullet. These features leave unique marks on bullets fired from a given weapon. Along with the double microscope for side-by-side comparison of bullets, invented at that same bureau, the helixometer made it possible to link a bullet from a crime scene to a specific weapon.

Souder's forensic work was not well known until 2014, when Kristen Frederick-Frost, curator of the NIST Museum, found a forgotten trove of Souder's old notebooks. She joined forces with John Butler at NIST, whose own work on DNA analysis has contributed to forensic science, and who compiled much of Souder's work from his notebooks.

Souder earned his physics PhD in 1916, from the University of Chicago. One of his teachers was Albert Michelson, who won the 1907 Nobel Prize for Physics for the precise interferometric measurements crucial to the 1887 Michelson–Morley experiment. Souder's PhD adviser was experimentalist Robert Millikan, who would earn the 1923 Nobel Prize for Physics, for research on the photoelectric effect and the charge on the electron. Souder published two papers with Millikan, and his dissertation about the photoelectric effect, in *Physical Review*.

Initially, Souder studied dental materials at NBS, to help the U. S. Army develop treatments for soldiers—a research award in dentistry is now named after him. But another pressing need soon arose, thanks to growing criminal activity in the 1920s. Much of this was fueled by Prohibition, the era from 1920 to 1933 when the U. S. banned alcoholic beverages, and criminal gangs fought viciously to control illegal bootlegging. Souder's notebooks show that he responded by providing forensic analysis of handwriting, typewriting, and bullets on more than 800 criminal cases for the Department of Justice, the Treasury Department, and other agencies. As the NIST researchers discovered, this resulted in an appreciative note from FBI director J. Edgar Hoover, and a gun carry permit for Souder that was justified protection for a witness in criminal trials.

These pioneering forensic approaches played roles in major cases. The historical research at NIST showed for the first time that Souder was involved in a sensational 1935 "trial of the century." It found Bruno Hauptmann guilty of kidnapping and killing the 20-month-old son of Charles Lindbergh, famous for the first solo flight across the Atlantic Ocean in 1927. Souder's study of the ransom notes in the case, with that of other handwriting experts, provided much of the evidence that put Hauptmann in the electric chair.

Weapons identification was likewise essential in another world-famous trial. In 1921, two Italian-born anarchists, Nicola Sacco and Bartolomeo Vanzetti, were convicted of shooting and killing two men during an armed robbery in Massachusetts. The verdict was widely condemned as having been unjustly influenced by the prevailing anti-radical sentiment in the U. S. At the final review of the case in 1927, Calvin Goddard, head of the Bureau of Forensic Ballistics, testified that the helixometer and the comparison microscope unequivocally showed that one fatal bullet and a cartridge case came from Sacco's pistol. Sacco and Vanzetti were executed, but controversy continued, although modern bullet analysis has confirmed Goddard's result.

These early forensic methods remain valuable, but forensic science in the U. S. has lost some of its luster. Reviews in 2009 and 2016 found that much of forensic practice has developed without the scientific rigor that would make it truly reliable in deciding guilt or innocence. The reviews called for improvements in forensic science, some of which are under way (see Ref. [1]). Souder, well-trained in scientific exactness, would have applauded these recommendations.

The NIST researchers found that in 1932 he was already calling for standards to be established for forensics equipment, for precise forensic data and its detailed recording, and for stringent testing to qualify forensics experts.

Souder was also well aware of the difficulties in presenting scientific evidence to judges and juries who lacked scientific training. He used oversized aluminum models of bullets to illustrate ballistic methods, and in 1954, writing in *Science*, described how to be an effective scientific witness in court. The article ended with Souder's rallying cry for the value of good forensic science that still resonates: "Justice is sometimes pictured as blindfolded. However, scientific evidence usually pierces the mask."

Reference

1. S. Perkowitz, *Phys. World*, Oct. 2019, 43–46.

Science, Fiction, and Art

Introduction

I expressed my long-standing interest in how science connects to literature, art, and the media in *Real Scientists Don't Wear Ties*. That has been affected by the covid pandemic. Like most of us, my interactions with art and film have been limited by the need to avoid public gatherings. Instead, online films and videos from varied sources have filled the gap. Among these, I viewed and wrote about independent and documentary science-based films more than in the past, some of which are significant in relating to current issues, from the pandemic itself to the impact of technology on people. Feature-length science fiction films are also represented in this section.

Science and art change each other

"The Art of Falling Fluid" (2014) shows how physical analysis of Jackson Pollock's method of applying paint to canvas supports his claim that he controlled the paint rather than letting it assume random shapes and gives the history of how he came to his particular approach. "Light in the Woods: Bruce Munro at the Atlanta Botanical

Science Sketches: The Universe from Different Angles
Sidney Perkowitz
Copyright © 2022 Jenny Stanford Publishing Pte. Ltd.
ISBN 978-981-4877-94-7 (Hardcover), 978-1-003-27496-4 (eBook)
www.jennystanford.com

Garden" (2015) describes an outdoor installation by the Australian artist Bruce Munro. The work updates Marcel Duchamp's famous "readymades" by using throwaway plastic water bottles, with optical fiber added to provide lighting effects. Both the bottles and the fiber are products of modern materials technology, which also produced the record-breaking black coating described in "Paint it Nanoblack" (2016). Its eerie other-worldly blackness has inspired artists.

Another connection between art and technology is that digital display methods have made the presentation of original artworks more satisfying. "Physics and Art in 2½ D" (2018) shows how computer-generated lighting of a high-resolution digital image of an original work allows a viewer to see the work's surface texture in quasi-3D on a flat device screen. "Altered States: 2D Digital Displays Become 3D Reality" (2019) describes how this "2½D" method can be extended to produce actual 3D simulations of artworks and museum pieces for the public to see and touch as an alternative to a fragile original.

Imagining the pandemic

Pandemics can elicit creative responses from artists, such as Albert Camus' famous novel *The Plague*. COVID-19 may not produce any such, but the Elia Kazan film *Panic in the Streets* (1950), about a modern resurgence of the disease that caused the Black Death in the 14th century, expresses concerns like those about COVID-19. As discussed in "Panic in the Streets: *Filming the Pandemic*," the story shows how the older disease can spread among people much as COVID-19 does. There are other parallels as well, such as the need for contact tracing, and the rejection of medical expertise to avoid facing the seriousness of the disease.

Another effort, the short story "The Machine Stops," published in 1909 by the English writer E. M. Forster, uncannily predicted the technology that has helped us break out of pandemic-induced isolation. As presented in "Only Disconnect!" (2020), the people in Forster's future world live underground and alone but interact widely through vision plates and have all of life's needs delivered by the central Machine. Forster warns the reader about the costs of abandoning nature and leaning too heavily on technology.

Documenting science onscreen

Any scientist can point to places where a science-based or science fiction film or television show displays bad science. Some of these presentations try to express the science correctly or at least plausibly, but documentaries and dramas made by independent filmmakers are typically more devoted to the science. The pieces under this heading provide a cross section of how both big and individual media sources deal with science.

"The Chemical Formula: Successfully Combining Chemistry, Science, and the Media" (2011) reports on a panel of media people and scientists (including me) whose discussion sheds light on the different factors that influence how science is presented onscreen. "Getting the Film Physics Right" (2020) focuses on films in the Labocine archive of science-based films from independent filmmakers; and from Sloan Science & Film (http://scienceandfilm. org/), a project of the Sloan Foundation that supports science films. Both articles highlight documentary and dramatic films that present physics and physicists, and related areas, in correct and meaningful ways. One film that grew out of Sloan support for a book, *Hidden Figures* (2016), went on to become a mainstream critical and commercial success.

"Facing Up to Facial Recognition" (2020) discusses three short films that cover different categories: a documentary about the neuroscience of recognizing faces; a science-based story about what happens when the male half of a romantic couple has a real-life neurological disorder, prosopagnosia, that prevents him from recognizing faces; and a science fiction story about the dire effects of equipping tiny lethal drones with facial recognition software and releasing them in swarms. "Sunday is Maroon: Synesthesia on Screen" (2020) also covers three films that span the categories, but all three are about synesthesia, the mixed sensory experience I discussed in the "The Power of Crossed Brain Wires."

"Science Cinema Online: The 13th Annual Imagine Science Film Festival" (2020) takes a broader view. It discusses a range of science documentaries and dramas presented online in 2020 as part of a film festival sponsored by Imagine Science Films, the partner organization of Labocine. Some are noteworthy for tackling specific scientific subjects correctly and in depth, such as *The Last Artifact*;

others for expressing important issues about technology in society, such as *Coded Bias*; and some for reminding us of historical examples, or projected future ones, where technology adversely affects people, such as *La Bobine 11004* (*Reel 11004*) and *FREYA*.

Fictional science and science fiction

Documentary films, the bulk of what was presented under "Documenting science onscreen," offer more or less straightforward views of existing science and scientists, seasoned with some dramatic license. Science fiction stories, however, take science or technology past their present limits. The extrapolation may be a likely one, as in many near-future stories; it may be purely speculative, as when humanity meets aliens; or it may violate known science, the classic example being faster-than-light travel. The pieces below are about feature-length science fiction films from Hollywood and elsewhere that illustrate these possibilities.

"Aliens: Love Them, Hate Them, or Relate to Them?" (2009) briefly surveys the history of alien creatures on screen since their appearance in the first science fiction film ever made, *Voyage dans la Lune* (*Voyage to the Moon*, 1902). Aliens have an especially interesting role in the South African film *District 9* (2009), where they carry implicit messages about apartheid and the enduring human reluctance to accept "others" outside our own groups. "Trapped on Mars" (2016) considers humanity's long-term preoccupation with our red neighbor, which has yielded dozens of science fiction films and TV productions. Among these, *The Martian* (2015) with Matt Damon stands out as a cinematic success that was also reasonably realistic about what we would really find on Mars.

"Science Advances and Science Fiction Keeps Up" (2018) treats two pairs of films: *The Day after Tomorrow* (2004) and *Geostorm* (2007), about climate change; and *Blade Runner* (1982) and *Blade Runner 2049* (2017) about the creation of replicants, artificial people. My comparison within each pair shows that even when films fail to get the science right, they are useful in following scientific trends and presenting them to the public. "Entropy and the End of the World in Tenet" reviews the film *Tenet*, released in August 2020. It is a time travel story with a novel approach to manipulating time that I explain in the piece.

The Art of Falling Fluid

When Jackson Pollock first dripped paint onto horizontal canvases to create works such as *Autumn Rhythm* (1950, MoMA), he became the founding spirit of a major art movement, Abstract Expressionism. Art historians have studied the intricate, looping patterns in Pollock's work ever since, but, on occasion, his paintings have also attracted attention from scientists.

One approach, originated by Richard Taylor, has been to subject Pollock's work to fractal analysis [1]. In 2011, however, a trio of researchers—physicist Andrzej Herczński, art historian Claude Cernuschi, and mathematician L. Mahadevan—examined the fluid dynamics of the paint itself. They studied how it behaved as Pollock dipped an implement into a can of paint, then wielded the tool to allow the paint to fall onto the canvas below. They concluded that Pollock controlled the paint as it descended under gravity, rather than letting it pour randomly. The process, the researchers suggest, should be called "streaming" rather than "dripping" because it produced continuous well-defined lines, not spatters.

These insights were given added meaning in 2012, when art historian Sandra Zetina and physicist Roberto Zenit of the National Autonomous University of Mexico (UNAM) reported a fluid dynamics study of another iconic artist, David Alfaro Siqueiros. A leader of the Mexican muralist movement, which carried great artistic and social weight from the 1920s to the 1940s, Siqueiros was one of the trio of great Mexican painters known as "Los tres grandes" (Diego Rivera and José Clemente Orozco were the others). Siqueiros was a true revolutionary: A devoted communist, in May 1940 he tried but failed to assassinate Stalin's enemy, Leon Trotsky, and he sought novel technology for art with the same revolutionary fervor. For example, in his painting *Collective Suicide* (1936, MoMA), Siqueiros used a quick-drying automotive paint developed by the Ford Motor Company to produce attractively complex abstract patterns. These patterns resulted from Siqueiros' "accidental painting" technique, in which he put a layer of paint atop another layer of a different color. The colors automatically interpenetrated to produce what Siqueiros called "the most magical fantasies and forms that the human mind can imagine."

Zetina and Zenit hypothesized that these forms arise from the Rayleigh–Taylor (RT) instability in fluids. Named after the British scientists Lord Rayleigh and Geoffrey Taylor, this phenomenon arises when a horizontal layer of fluid supports another layer of denser fluid. This situation is inherently unstable, like a big boulder insecurely perched on a small rock. The slightest irregularity at the interface of the two fluids will cause gravity to pull the denser fluid into the less dense one. This produces complex turbulent extrusions that take on fantastic shapes in both fluids.

To test the theory, Zetina and Zenit simulated Siqueiros' method in a controlled manner by putting a layer of white paint on a layer of less dense black paint. As their video (http://ow.ly/sGn6S) of the resulting evolution shows, what emerges is a beautiful, unpredictable pattern that resembles those in *Collective Suicide* and obeys the mathematics of RT instability.

What does this have to do with Jackson Pollock? Remarkably, despite their different national origins, Pollock encountered Siqueiros and his methods early in his career. In the same year that Siqueiros created *Collective Suicide*, he also ran an experimental workshop in New York to introduce his novel methods to other artists, including a 24-year-old Pollock. The attendees probably saw *Collective Suicide*, but during the workshop, they did more than just experiment with RT instability. As one of them wrote, "We sprayed [paint]...used it in thin glazes or built it up into thick globs...we poured it, dripped it, spattered it, hurled it at the picture surface."

According to Helen Harrison, director of the Pollock–Krasner House and Study Center in East Hampton, New York, this early exposure to paint in motion was important in developing Pollock's technique. In her book *Jackson Pollock* (Phaidon Press, 2014), Harrison notes that Pollock knew of Siqueiros' unconventional approaches before the 1936 workshop. Nevertheless, she observes, Pollock's brother Charles put great weight on the influence of that workshop, which he considered "a key experience in Jackson's development" that "must have stuck in his mind to be recalled later, even if unconsciously, in evolving his mature painting style."

Physical analysis illuminates the ways that both artists used a natural effect—fluids falling under gravity—to produce their works. However, it does not explain everything, since the artists also made creative choices. Siqueiros declared, correctly, that his patterns were made by the paint itself, and so accepted the resulting unpredictability. Pollock, on the other hand, based his approach on apparently haphazard but actually well-controlled streams of paint falling through air. As he put it, "I can control the flow of paint: there is no accident." And by itself, the physics of fluids cannot answer another important question: why are we pleased by Siqueiros' turbulent patterns and Pollock's looping ones? The answer may have to wait until we can complement physical analysis with a true neuroscience of aesthetics—if this ever becomes possible.

Reference

1. R. Taylor, *Phys. World*, Sept. 2013, 37–41.

Light in the Woods: Bruce Munro at the Atlanta Botanical Garden

A hundred years ago, Marcel Duchamp hugely influenced painting and sculpture when he showed that unexpected things can be turned into art. He did this by enlarging the idea of a "found object," which he called a "readymade." This is any ordinary object that is displayed as art and so becomes art. Duchamp's readymades, such as a glass vial full of Parisian air and a porcelain urinal, are now seen as helping to define 20th-century art.

Today, as plastic replaces glass and porcelain in our throwaway society, artists like Bruce Munro can draw on new types of readymades. Munro combines appreciation of light in nature with the use of readymades in his exhibition *Light in the Garden*, installed last May at six sites in the Atlanta Botanical Garden and remaining there until October.

Munro was born in England but also has roots in Australia, which has influenced his distinctive artistry. After earning an art degree in the U. K., he moved to Sydney, where his job was to make display signs using a plastic that glows under ultraviolet light. Though this experience plays into his art, it could have happened anywhere. What's more important is that Munro had a uniquely Australian moment when in 1992 he visited the great natural feature called Uluru (also known as Ayers Rock). Rising out of the desert in the midst of the outback, this enormous sandstone monolith glows red at dawn and sunset and stands out even more than Georgia's Stone Mountain. It has spiritual meaning too, for it is sacred to the local aborigines.

Munro, who comes across in person as both solidly grounded and wide open to any kind of inspiration, felt what he calls an "ever present zing of something special" at Uluru. The site energized him, he writes, with a vision of life springing out of the desert, as when rain makes the arid environment suddenly blossom. In 2004, he expressed this vision in the installation "Field of Light." Its theme was light interacting with nature, shown by myriads of flower-like vertical stems holding glowing globes of light and placed in a large field near Munro's home in England.

Since then, Munro continues to show light within nature in different large-scale outdoor installations, including several in the U. S. "Forest of Light," his latest version of this signature work within *Light in the Garden,* occupies the Storza Woods, a hardwood forest at one end of the Garden. The work consists of tens of thousands of light sources on stems scattered throughout the forest floor. They are interconnected via optical fiber, thin glass conduits that carry light of any color to any desired location, and they can be viewed at ground level or from the elevated Canopy Walk that winds through the Woods. Peering down as twilight and then darkness settle in, you see one set of lights, then another and another, start to glow, each in its own muted color. The process and the final effect are natural, beautiful, and slightly disorienting, like an enhanced reflection of the constellation of lights in the night sky.

Other elements in the exhibition are more sculptural. Placed in indoor Garden venues are "Three Degrees," a set of three sinuous shapes reminiscent, in Munro's telling, of curvaceous female forms; and "Eden Blooms," fanciful alien-looking floral forms with bright colors that might have come from some distant tropical planet. "Swing Low" is an outdoors arrangement of light-filled spheres suspended over water.

The remaining two, "Beacon" and "Water-Towers," are where Munro uses his readymades, which could not be more ordinary and ubiquitous. They are transparent plastic water bottles, like those you buy filled with purified water at your local convenience store or supermarket. Incredibly, some 50 billion of these were used in the U. S. last year.

When I asked Munro if he was thinking of Duchamp when he decided to use the bottles, he replied that it was a pragmatic choice, made because they were cheap. But, he adds, the decision was also based on "one of those fortunate moments when I spotted the beauty of a stack of bottles in a retail store...the work of Duchamp, Picasso et al. had seeped into my subconscious."

Munro creates "bottled light" by filling the water bottles with the same kind of optical fiber as in "Forest of Light." The colored light the fiber carries diffuses through the plastic to give the whole bottle a rich glow. In "Water-Towers," bottles are stacked in 20 separate 6-feet-tall cylinders, placed around the Garden's Aquatic Plant Pond. At night, each column is luminous with a single color that changes as the cylinders cycle through different shades. The effect is of a hi-tech update of some ancient columned structure like a Greek temple.

"Beacon," located near the Garden's Great Lawn, works in both daylight and darkness. Its framework is made of girders forming a big geodesic dome 15 feet across. The open triangular spaces in the geodesic pattern are filled with nearly 3000 water bottles pointing inward and again containing optical fiber. During the day, you can look through the bottles to see the intricate internal engineering, and if you step back, the structure resembles an enormous, faceted diamond. At night, when varied jewel tones piped in by the optical fiber are visible, it becomes any gem you want it to be or the best kaleidoscope ever.

Munro's artistic use of light has appeared in many venues from gardens to cathedrals in the U. S. and the U. K. Asked about his Atlanta exhibition, he says it was a privilege to help inaugurate the next stage in the development of the Storza Woods in May, and that he especially liked its unique Canopy Walk. But among all his varied artistic experiences, Australia still exerts a special pull.

Munro finds inspiration in the offbeat film *The Last Wave* (1977) by the Australian director Peter Weir, an unsettling story based on aboriginal mythology about impending global catastrophe. Expressing this "dream world" in light is a future project for the artist. Then there is his continuing engagement with Uluru, which he has just revisited in hopes of celebrating his 1992 revelation with a new "Field of Light." Creating a light installation at Uluru, Munro says, "has been a 23-year journey so it's essential I reduce the time lag between concept and execution! Getting older makes time a very precious commodity!"

Fortunately, Bruce Munro, Uluru, and even plastic readymades— should he need them—will be with us for a long time to come.

Paint It Nanoblack

Ever since our ancestors painted images on the walls of caves, artists have sought pigments to represent the 10 million tints that humans can differentiate. Some artists, however, choose to portray the world in black and white, a radical simplification that pulls us in with surprising force. Over the centuries, these artists have employed substances from India ink to titanium white to portray shadow, light, and contrast. Now they have a new ally: researchers who are using optical design principles, nanotechnology, and inspiration from nature to create deeper blacks and purer whites.

Black coatings and materials that absorb photons at visible and longer wavelengths have long been used to reduce unwanted stray light within optical equipment such as cameras and telescopes. These materials also have uses in military stealth operations and efficient solar energy conversion. Many are based on paints that include carbon or graphite, and they typically reflect only a few percent of incoming light while absorbing the rest. But now we have graphene, a 2D form of carbon that can be rolled up into nanotubes to offer a way to create even deeper blacks.

Since 2007 scientists have competed in a "blackest black" arms race, developing methods to create nanoforests of aligned carbon nanotubes that both scatter and absorb incoming light [1]. One of the resulting products, Vantablack, reflects only 0.035% of visible light, and looks qualitatively different from "normal" blacks even to the naked eye. A display at London's Science Museum in February 2016 showed how Vantablack's negligible reflectivity obliterates an object's surface features, creating the eerie look of a hole in space or a tunnel to another dimension.

Artists quickly grasped the value of this darkest black. *National Geographic* photographer Peter Essick, known for his striking black-and-white images, notes that current digital printers do not reproduce the full black-to-white range that digital cameras capture. He speculates that Vantablack, or something like it, will soon be incorporated into digital printer ink. Vantablack has also inspired the prize-winning U. K.-based sculptor Anish Kapoor. In a December 2015 talk at the Hirshhorn Museum in Washington, DC, Kapoor spoke about creating a huge hollow steel construction coated internally with the material. This would, he said, give viewers inside it the sense that "what's inside the object is bigger than what's outside. That's how we are. What's inside us has a completely bigger imaginative reality."

Vantablack is expensive and hard to make, and its military potential makes it subject to export controls. The main restriction on its broad use, though, is that its creator, Surrey NanoSystems, has given Kapoor exclusive artistic rights to the material, angering artists who spoke against arbitrarily limiting access to an artistic medium.

Fortunately, nature's own nanoscale experiments provide alternative paths to better blacks and whites. Take the black areas seen on butterfly wings. Here, scales containing the pigment melanin absorb sunlight to regulate the insect's temperature. Pete Vukusic and colleagues at the University of Exeter, U. K., have shown that in the *Papilio ulysses* butterfly, such absorption is augmented by a nanometer-sized fine structure within the scales, a honeycomb-like network of ridges and struts made of chitin, a natural polymer. This structure gives light a longer optical path through the melanin and increases absorption from 52% for the pigment alone to 95% at a wavelength of 600 nm. In China, Qibin Zhao of Shanghai Jiao Tong University and co-workers showed that nanoscale enhancement makes carbon itself a better absorber. When a sheet of amorphous carbon is patterned with a ridged nanostructure that is found in a different butterfly, the carbon becomes blacker as its reflectivity drops from over 12% to under 1%.

White compounds can also be complicated to create. Unless they are "true whites," with the same reflectivity at all visible wavelengths, they carry a subtle tinge that the human eye can perceive, which affects calibration standards and paper made for printing and artistic use. But here, too, natural nanostructures produce whites that provide models for human-made versions. One example is found in *Cyphochilus*, a genus of beetle whose white body is thought to provide protective coloration in environments containing white fungi. As determined by Vukusic, Lorenzo Cortese at the University of Firenze in Italy and others, its whiteness comes from highly efficient scattering within chitin scales 250 μm long that contain random networks of filaments 250 nm across. This structure amounts to a natural photonic solid optimized to produce an exceptionally bright true white with nearly constant reflectivity, from a layer only 5 μm thick. This is far thinner than synthetic layers that produce a comparable white. With its desirable optical and physical properties, this natural geometry could be used for ultra-white coatings, enhanced LED sources, and for camouflage.

Nanotechnology is still new. We are barely beginning to exploit it and to use lessons from nature to create novel materials. It may not be long before a digital printer uses ink derived from butterflies to produce exceptional images on paper derived from beetles, or before artists everywhere can freely extend their imaginations with the blackest blacks and whitest whites.

Reference

1. J. Cartwright, *Phys. World*, Nov. 2015, 25–28.

Physics and Art in 2½D

We live in three spatial dimensions—a good thing, since 2D and 1D are less interesting than 3D. The only problem is in knowing how to artistically portray the 3D world on 2D paper or canvas. No one understood how to do this until about 1415, when the Renaissance architect Filippo Brunelleschi invented linear perspective to display 3D scenes on 2D surfaces. But even "flat" artworks extend slightly into 3D. The underlying surface may impart texture, and artists create a complex coated surface when they build up layers of pigment on their images.

This texture is a material record of the artist's effort that makes an original work unique but does not show up well in ordinary photos and screen images. Now, though, applied optics and sophisticated computation make it possible to directly experience the surface topography of an artwork on a flat display screen. Loïc Baboulaz, an imaging scientist at the Swiss Federal Institute of Technology (EPFL) in Lausanne, has led a project since 2012 to improve the digital display of artworks.

Digital display, in general, has made huge strides in applications such as computer-generated imagery (CGI), which creates fantastic, but seemingly real scenes in films. This is done by developing the "bidirectional scattering distribution function" (BSDF) for every point in a scene. The BSDF shows how reflection and scattering affect a light ray entering from an arbitrary direction, to produce the outgoing ray that enters a viewer's eye. The function can even show how the scene would look if the light source is shifted.

In CGI, the BSDF is derived from optical theory, from mathematical models of the shapes, and from the optical properties of objects in the imaginary scene. The EPFL scientists, however, wanted a BSDF based wholly on actual data from an artwork. In 2014 they showed how to construct the light transport matrix (LTM), a digitized BSDF, for a real nearly planar object—in this case, a pane of stained glass. They illuminated it from different directions and recorded the results with a high-resolution camera. In principle this process picks up all the object's interactions with light through its surface, as well as subsurface reflections, scattering, and refraction (including irregularities). This could create an LTM with a staggering number of elements. But in practice, a flat object with shallow surface topography generates a sparse matrix with relatively few elements, hugely reducing the computational power needed to manage the data.

This approach is used by Artmyn, a startup in Saint-Sulpice, Switzerland, co-founded by Baboulaz, who is now its chief technology officer. Artmyn produces "2½D" reproductions of artworks by illuminating them with light-emitting diodes at different visible and ultraviolet wavelengths, generating thousands of images at high resolution and color fidelity. Its proprietary software turns up to a terabyte of data into an image that can be displayed and manipulated in real time, on a computer or mobile device. The key feature is that the lighting can be varied to produce 3D-like highlighted peaks and shadowed valleys in the surface topography.

To illustrate exactly what this technology can do, Artmyn collaborated with Sotheby's auction house to digitize Marc Chagall's *Le Printemps*, a 1975, 46 × 55 cm oil painting of a couple embracing, flowers, a donkey, and a village in shades of red, green, and blue. The interactive online display can be adjusted such that you can view everything from the full painting, down to every fleck of color Chagall applied as it clings to the weave of the canvas—illustrating that, as Picasso noted, Chagall truly "understands what colour really is." Other Artmyn displays, from the Bodmer Foundation in Switzerland, show a Sumerian clay tablet circa 2000 BC covered with cuneiform script, and Papyrus 66, an Egyptian manuscript of the New Testament from 200–300 AD. Changing the lighting emphasizes the symbols impressed into both artifacts and shows cracks and variations that give a sense of their materials and their great ages.

Surprisingly, another digitization from Bodmer is a physics image—page 48 of Isaac Newton's original *Philosophiæ Naturalis Principia Mathematica* (1687) with text and a diagram related to centripetal force. Zooming in and changing the light direction show faint reversed images of what is printed on the underside of the page, its texture, and distortions in the paper that indicate its age and use. This particular copy belonged to Newton's contemporary Gottfried Leibniz, who independently invented calculus. His cramped but clear notes in the margin hint at his presence, along with several reddish spots. Leibniz was known to read while smoking in bed, and the marks are burns left by cinders from his pipe.

For all the pleasure these images give to art lovers and scholars, the owners of the original works need to know that the Artmyn process is safe. To reassure them, Artmyn measured the temperature of artworks while being scanned and found no increases. Even then, Baboulaz told me, some owners are not ready to share their digitized art with the world. I wonder what will happen if, in the future, the stored digital data are used to create exact physical copies via 3D printing. Then we would have to ask if copies that are identical down to microns would devalue the magic of the work that felt the artist's touch, and if owning an "original" would take on a whole new meaning.

Altered States: 2D Digital Displays Become 3D Reality

It's a natural impulse to reach out and touch an original artwork, perhaps to feel the strong brushstrokes in van Gogh's *Starry Night* or to trace the shape of a compelling sculpture. You can't though, and for good reason: A multitude of individual touches would soon damage the work, so museums sensibly use "Please don't touch" signs, velvet ropes, and alert guards to keep viewers at a distance. It helps that those same museums have put their collections online so you can appreciate great art right on your digital device. However, even at high resolution, images on flat screens do not clearly show surface texture or convey volumes in space. But now researchers in art and technology are finding ways for viewers to experience the texture of artworks in 2D and the solidity of those in 3D.

The missing third dimension is significant even for flat works, which typically show the texture of the background paper or canvas, or of the pigment. Some nominally 2D works are inherently textured, such as Canadian artist Brian Jungen's *People's Flag* (2006), an immense vertical hanging piece made of red textiles. Helen Fielding of the University of Western Ontario has perceptively noted how vision and touch intertwine in this work:

> As my eyes run across the texture of the flag, I can almost feel the textures of the materials I see; my hands know the softness of wool, the smoothness of vinyl. Though touching the work is prohibited...my hands are drawn to the fabrics, subtly reversing the priority of vision over touch...

Textural features like these are a material record of the artist's effort that enhances a viewer's interaction with the work. Such flat but textured works are art in "2.5D" because they extend only slightly into the third dimension. Now artworks shown in 2.5D and 3D on flat screens and as solid 3D models are giving new pleasures and insights to art lovers, curators, and scholars. As exact copies, these replicas can also help conserve fragile works while raising questions about the meaning of original art.

One approach, developed at the Swiss Federal Institute of Technology (EPFL) in Lausanne, creates a digital 2.5D image of an artwork by manipulating its lighting. Near sunset, when the Sun's rays enter a scene at an angle, small surface elevations cast long shadows that make them stand out. Similarly, the EPFL process shines a simulated light source onto a digital image. As the source is moved, it produces highlights and shadows that enhance surface details to produce a quasi-3D appearance.

This approach has links to CGI, computer-generated imagery, the technology that creates imaginary scenes and characters in science fiction and fantasy films. One powerful CGI tool is an algorithm called the bidirectional scattering distribution function (BSDF). For every point in an imagined scene, the BSDF shows how incoming light traveling in any direction would be reflected or transmitted to produce the outgoing ray seen by a viewer. The result fully describes the scene for any location of the light source.

In films, the BSDF is obtained from optical theory and the properties of the imaginary scene. The EPFL group, however, generated it from real art. In 2014, they illuminated a pane of stained glass with light from different directions and recorded the results with a high-resolution camera, creating a BSDF and showing that the method works for nearly planar items. This approach has been commercialized by Artmyn, a Swiss company co-founded by Loïc Baboulaz who led the EPFL team. Artmyn makes 2.5D digital images of artworks by lighting them with LEDs at different visible wavelengths to provide color fidelity, and at infrared and ultraviolet wavelengths to further probe the surface. The result is a BSDF with up to a terabyte of data.

As an illustration, Artmyn has worked with Sotheby's auction house to digitize two Marc Chagall works: *Le Printemps* (1975, oil on canvas), a village scene with a couple embracing, and *Dans L'Atelier* (1980, tempera on board), an artist's studio. The Artmyn software lets a viewer zoom from the full artwork down to the fine scale of the weave of the canvas, while moving the lighting to display blobs, islands, and layers of pigment. This reveals how Chagall achieves his effects and clearly illustrates the difference between oils and tempera as artistic media. Currently in process for similar digitization, Baboulaz told me, are a Leonardo da Vinci painting and a drawing, in recognition of the 500th anniversary of his death this year.

Artmyn has also digitized cultural artifacts such as a Sumerian clay tablet circa 2000 BCE covered in cuneiform script; signatures and letters from important figures in the American Revolution; and a digital milestone, the original Apple-1 computer motherboard. These 2.5D images display signs of wear and of their creator's presence that hugely enhance a viewer's visceral appreciation of the real objects and their history.

For the next step, creating full 3D representations and physical replicas, the necessary data must be obtained without touching the original. One approach is LIDAR (light detection and ranging), where a laser beam is scanned over the object and reflected back to a sensor. The distance from the laser to each point on the object's surface is found from the speed of light and its travel time, giving a map of the surface topography. LIDAR is most suitable for big artifacts such as a building façade at a coarse resolution of millimeters. Other approaches yield finer detail. In the triangulation method, for instance, a laser puts a dot of light on the object while a nearby camera records the dot's location, giving data accurate to within 100 μm (0.1 mm). Copyists typically combine scanning methods to obtain the best surface replication and color rendition.

One big 3D copying effort is underway at the Smithsonian Institution, whose 19 museums preserve 155 million cultural and historic artifacts and artworks. Since 2013, the Smithsonian has put over 100 of these online as interactive 3D displays that can be viewed from different angles, and as data for 3D printers so people can make their own copies. The objects, chosen for popularity and diversity, include the original 1903 Wright Brothers flyer; a highly decorated 2nd century BCE Chinese incense burner; costume boots from the Broadway musical *The Wiz* from 1975; a mask of Abraham Lincoln's face from shortly before his assassination in 1865; and for the 50th anniversary of the Apollo 11 moon landing, astronaut Neil Armstrong's spacesuit. Recently added is a small 3D version of a full-sized dinosaur skeleton display at the National Museum of Natural History showing a T-rex attacking a triceratops, for which hundreds of bones were scanned by LIDAR and other methods.

A different goal animates the 3D art and technology studio Factum Arte in Madrid, Spain. Founded by British artist Adam Lowe in 2001, Factum Arte protects cultural artifacts by copying them, using its own high-resolution 3D scanning, printing, and fabrication techniques.

Museums already use copies to preserve sensitive artworks on paper that need long recovery times in darkness and low humidity between showings. During these rests, the museum displays instead high-quality reproductions (and informs patrons that they are doing so). In a recent interview entitled "Datareality," Adam Lowe expressed his similar belief that an artistically valid copy can provide a meaningful viewing experience while preserving a fragile original. One of his current projects is to replicate the tombs of the pharaohs Tutankhamun (King Tut) and Seti I, and queen Nefertari, in the Egyptian Valley of the Kings. The tombs were sealed by their builders, but once opened, they are deteriorating due to the throngs of visitors. As Lowe recently explained, "by going to see something that was designed to last for eternity, but never to be visited, you're contributing to its destruction."

The copies, approved by Egypt's Supreme Council of Antiquities, will give visitors alternate sites to enter and view. At a resolution of 0.1 millimeter, the copies provide exact reproductions of the intricate colored images and text adorning thousands of square meters in the tombs. The first copy, King Tut's burial chamber, was opened to the public in 2014, and in 2018, Factum Arte displayed its copied "Hall of Beauties" from the tomb of Seti I.

Earlier, Factum Arte had copied the huge Paolo Veronese oil on canvas *The Wedding Feast at Cana* (1563, 6.8 m × 9.9 m), which shows the biblical story where Jesus changes water into wine. The original was plundered from its church in Venice by Napoleon's troops in 1797 and now hangs in the Louvre. The full-size copy, however, commissioned by the Louvre and an Italian foundation, was hung back at the original church site in 2007.

Factum Arte's efforts highlight the questions that arise as exact physical copies of original art become available. Museums, after all, trade in authenticity. They offer viewers the chance to stand in the presence of a work that once felt the actual hands of its creator. But if the copy is indistinguishable from the work, does that dispel what the German cultural critic Walter Benjamin calls the "aura" of the original? In his influential 1935 essay *The Work of Art in the Age of Mechanical Reproduction*, he asserted that a copy lacks this aura:

> In even the most perfect reproduction, one thing is lacking: the here and now of the work of art – its unique existence in a particular place. It is this unique existence – and nothing else – that bears the mark of the history to which the work has been subject.

The Factum Arte reproductions show that "original versus copy" is more nuanced than Benjamin indicates. The Egyptian authorities will charge a higher fee to enter the original tombs and a lower one for the copies, giving visitors the chance to feel the experience without causing damage. Surely this helps preserve a "unique existence in a particular place" for the original work. And for the repatriated *Wedding at Cana*, Lowe tellingly points out that a copy can bring its own authenticity of history and place:

> Many people started to question about whether the experience of seeing [the copy] in its correct setting, with the correct light, in dialogue with this building that it was painted for, is actually more authentic than the experience of seeing the original in the Louvre.

We are only beginning to grasp what it means to have near-perfect copies of artworks, far beyond what Walter Benjamin could have imagined. One lesson is that such a copy can enhance an original rather than diminish it, by preserving it, and by recovering or extending its meaning.

Copying art by technical means has often been an unpopular idea. Two centuries ago, the English artist William Blake, known for his unique personal vision, expressed his dislike of mechanical reproduction such as imposing a grid to copy an artwork square by square. Current technology can also often stand rightfully accused of replacing the human and the intuitive with the robotic and the soulless. But properly used, today's high-tech replications show that technology can also enlarge the power and beauty of an innately human impulse, the need to make art.

Panic in the Streets:
Filming the Pandemic

In the history of film, there are depictions of dystopias, such as the world after climate change, giving viewers a safe way to consider potentially awful outcomes for humanity. As we worry about the global crisis of COVID-19 and its unknown future, it's natural to seek movies that offer insight into pandemics. The list of such films can begin with the classic horror story *Nosferatu* (1922), where the vampire Count Orlok, together with rats carrying a plague, bring death to a small, 19th-century German town. Fleas from rats spread the real disease called plague, the cause of the Black Death that killed much of the European populace in the 1300s. Three decades after *Nosferatu*, the 1950 film *Panic in the Streets* brings plague to a big 20th-century American city. Plague is not the coronavirus, yet this film shows their shared features and can illuminate issues with the response to our current viral pandemic.

Panic in the Streets was director Elia Kazan's sixth feature film, sandwiched between his better known, Oscar-winning landmarks *Gentleman's Agreement* (1947), *A Street Car Named Desire* (1951), and *On the Waterfront* (1954)—although it too won an Oscar for Best Motion Picture Story. Shot in black-and-white *noir* style, on location in New Orleans, *Panic in the Streets* begins with a poker game near the city's docks. The big winner is Kochak, newly arrived in the city, but he is ill with flu-like symptoms and soon leaves with his winnings. The other players are gangsters, and their leader Blackie (Jack Palance in his screen debut) and two of his hoodlums follow Kochak. After Blackie shoots and kills Kochak and takes the money, his henchmen dispose of the body.

When the body is found and the police surgeon examines it, he sees something that makes him call in Dr. Clinton Reed (Richard Widmark) of the uniformed U. S. Public Health Service. Reed finds that the dead man was sick with pneumonic plague, which is easily transmitted between people. Immediately worried that the disease may spread to others in the population, Reed orders the body cremated and inoculates everyone who has been near it with the antibiotic streptomycin.

Despite these measures, Reed fears that the murderer and any accomplices are infected and must be found in order to fully control the disease's outbreak. As he forcefully tells the city's mayor and other officials, plague is a serious matter. It caused the Black Death and can still bite. In 1924, 26 people in Los Angeles died of pneumonic plague, the most dangerous form that can spread person-to-person "like the common cold...on the breath, sneezes or sputum of the sick." "I may be an alarmist," adds Reed, but if plague "ever gets loose it can spread over the entire country." Plague develops and spreads rapidly, so police have only 48 hours to solve the crime. To deflect blame over not yet doing this, the police chief insists that the victim died by gunshot, not from illness, but then reluctantly assigns police captain Tom Warren (Paul Douglas) to find the murderer.

Figure 3 Elia Kazan's *Panic in the Streets* (1950) foresaw the medical and human issues arising in today's coronavirus pandemic. Here U. S. Public Health Service doctor Clinton Reed (Richard Widmark) injects an antibiotic to combat pneumonic plague, which spreads much as COVID-19 does.

Dr. Reed's comments about pneumonic plague could come right from current headlines about COVID-19 and how it is transmitted. Bubonic plague swells the lymph glands, and septicemic plague blackens the skin (hence the Black Death), causing high death rates; but pneumonic plague attacks the lungs and is the only form that spreads directly between people. It is 100% fatal if not treated quickly, justifying Reed's urgency in finding infected individuals. It is true too that plague is not gone, even today. Hundreds of cases appear worldwide every year, with occasional clusters as in 1924 Los Angeles, a real event. Fortunately, plague does not come from a virus as does COVID-19, but from the bacterium *Yersinia pestis*, and can be treated with antibiotics. Reed's use of streptomycin is valid but would be worthless against the coronavirus.

The reaction to Dr. Reed and his message as he tries to persuade others about the incipient pandemic also has contemporary echoes. The mayor accepts Reed's expertise, but others are skeptical, or like the police chief, they act first to defend their own interests. At one point, Reed has to pull rank, forcing the local police to accept preventive injections by threatening to impose quarantine. Captain Warren initially accuses Reed of furthering his own career, but when he sees that Reed really is committed to preventing a medical disaster, he comes to trust the doctor's integrity and joins forces with him.

Today's headlines similarly show that some of those in charge blame the messenger who brings bad medical news or are more interested in protecting their own assets than in public health. We also see the tension between broad compliance with pandemic safety guidelines from Federal scientists, and pushback from individuals and local governments unwilling to trust the experts and join a collective response. But we also see selfless actions from medical personnel who, like Dr. Reed, work to save lives even under personal risk. It's hard not to conclude that a pandemic is a stress test that exposes the extremes of human nature.

Panic in the Streets depicts the reality that disease does not stop at borders. Even in 1950, when air travel was less extensive than it is today, Reed understands how quickly plague could spread around the world. A port city like New Orleans, where the film is set, is especially vulnerable as we learn when Reed's investigation uncovers the identity and history of the dead patient zero. Kochak reached the port as a stowaway—carrying both smuggled goods and the illness—aboard a freighter from Oran (significantly, the Algerian city where Albert Camus set his great 1947 novel *The Plague* about a deadly infestation). With these clues, Warren and Reed find Blackie and his gang members, bring them to justice, and identify the contacts they and Kochak made.

While *Panic in the Streets* is a gripping fictional story, the science is real and accurate. The blend of story and science deepens the film's lessons about pandemics and our response to them. Another lesson comes from the film's focus on plague, a centuries-old disease that subsides but never completely dies. Whatever the source, once a pandemic is widely rooted in the environment, no one can guarantee that it will never reappear. Vigilance is essential.

We must also absorb the message in the film's final scene. As a tired Dr. Clinton Reed returns to his family, he hears on the radio that "all contacts have been found and inoculated." We won't hear that welcome announcement about COVID-19 until testing and contact tracing reach widespread levels and we have a vaccine.* Otherwise, we may have our own real panic in the streets.

---

*This piece was written before effective vaccines were developed. Unfortunately, even with these vaccines, we cannot yet say that we have defeated the virus.

Only Disconnect!

Chances are, you're hunkered down at home right now, as I am, worried about COVID-19 and coping by means of Instacart deliveries, Zoom chats, and Netflix movies, while avoiding others and the outside world. As shown by the experience in China [1] and recent studies [2], voluntary isolation to extreme lockdown are effective in slowing or stopping pandemics. But as cabin fever sets in and we miss friends and family, it's natural to wonder about the mental and social cost of widespread physical separation. Yet the surprising fact is your tech-heavy exile is not a brand-new idea in the Internet age. It was foreseen long ago. The prophet was the great English novelist E. M. Forster. His literary classics *Howards End* (1910), about British social relationships in Edwardian times, and *A Passage to India* (1924), about British rule in India, are also known through their masterly film versions.

Howards End is the origin of Forster's famous catchphrase, "Only connect!," expressing his belief in the essential need for human relationships. Protecting those connections in an increasingly mechanized world was the theme of Forster's 1909 story, "The Machine Stops." The science fiction story is a protest against what Forster saw as the dehumanizing effects of technology. It is meant to be a counterweight to H. G. Wells' faith in the value of scientific and technological progress. Forster was firmly on the humanities side of the Two Cultures, the other being science, delineated by another English novelist, C. P. Snow, in the 1950s. But time and progress have a funny way of reshaping literature. Today the "The Machine Stops" can be read as a remarkably prescient depiction of the Internet. What's more, Forster might be astonished to learn his Machine can draw us together and preserve our humanity and relationships. That's not say, however, that a warning about technology in "The Machine Stops" doesn't linger.

"The Machine Stops" is set in an undefined future era when people have left the Earth's surface, perhaps after a catastrophe such as climate upheaval or a deadly pandemic, and live underground, each in a separate room. We meet Vashti, a middle-aged member of this society, in her personal cell that she rarely leaves. Lacking sunlight or exercise, humanity has physically deteriorated. Vashti is described as "a swaddled lump...five feet high, with a face as white as a fungus." Her small room contains only a desk and an armchair but offers amenities. At the press of a button, Vashti can summon illumination, food, clothing, music and literature, even a warm or cold bath. If she is ill, medical equipment remotely guided by a physician diagnoses and treats her. For companionship, she uses a communications device, a hand-held blue video plate not much different from a smartphone set up for video chat.

All this is made possible by a global entity, the Machine, which supplies everything Vashti and anyone else could want. The Machine functions like a combination of the systems and processes that help us endure today's enforced isolation: the Web, the Internet and Internet of Things, online ordering and delivery, and telemedicine. Just as we are in today's social media, Vashti is also linked to many others through the Machine. She knew "several thousand people, in certain directions human intercourse had advanced enormously," writes Forster. She interacts with these one-on-one through her blue video plate or addresses a large audience in their separate rooms about her studies in the history of music. As Forster relates, "the clumsy system of public gatherings had long since been abandoned."

This limited life based on remote human contact and mediated by the Machine seems to fulfill Vashti's physical, emotional, and intellectual needs. With her fellow humans, she worships the Machine with religious fervor. But some in this society resent the Machine and feel a great lack in their world—among them, Vashti's adult son Kuno.

Like every other child in this society, Kuno was raised in a public nursery. He now lives far from his mother on the other side of the Earth, but Vashti still feels connected to him. When he calls on the blue plate and asks her to visit him because he has something to say "not through the wearisome Machine," she cannot say no, although the trip is difficult for her. Leaving her cell to board an airship, interacting face-to-face and physically with an air hostess and a few other travelers, and looking out on the natural Earth as she flies, are all deeply upsetting.

What Kuno has to tell Vashti is even more unsettling. He has long felt the need for more space, more freedom than life under the Machine provides. His yearning leads him finally, and illegally, to break out and reach the Earth's surface through an old railway tunnel. There he rediscovers the natural world of Sun and stars, hills and clouds, and grass. But the Machine had noted his absence and sent long worm-like tentacles that drag him back to his underground cell. Now he is faced with homelessness, meaning permanent expulsion to the surface.

This is nearly the end of Forster's story. Its last scene shows a greater cataclysm than homelessness for Kuno, as Forster gives his final opinion about the foolishness of relying too much on technology. He imagines that the Machine slowly and then quickly deteriorates. Small issues become big ones until suddenly the Machine completely stops. Humanity is utterly unprepared for this and perishes underground, the only hope for the race a few souls who may survive under primitive conditions on the Earth's surface.

When we see the Machine as more than the Internet, Forster's message comes into view. Today we face the Machine grinding to a halt—the global machine of public health, of governmental power and its ability to reassure us, of business and trade. Kuno's story brings the final warning. Our screens show only pixels, not the full experience of the natural world and real people. We're fortunate to have this technology as a backup; but once the present emergency ends, our fondest wish is to again see, touch, and talk to each other directly, not only through the wearisome Machine.

References

1. Cyranoski, D. What China's coronavirus response can teach the rest of the world. Nature.com (2020).

2. Mahtani, K. R., Heneghan, C., and Aronson, J. K. What is the evidence for social distancing during global pandemics? The Centre for Evidence-Based Medicine (2020).

The Chemical Formula: Successfully Combining Chemistry, Science, and the Media

It's hard to know what some 500 chemists were expecting when they filed into a ballroom for an event called *Hollywood Chemistry* this past March 27, at the big annual meeting of the American Chemical Society (ACS) in Anaheim, California. What they got may have been a surprise: a 2-hour session that indeed covered some traditional chemistry, but mainly presented more drama, special effects, and laughs than the standard ACS scientific session, while making important points about chemistry and science in the media.

That's because *Hollywood Chemistry* featured a panel that combined writers for popular science- or medicine-focused TV series, with scientists who deal with science as it is treated in TV and film. The panelists showed clips, which provided the drama and special effects; spoke entertainingly about the tribulations of merging science with entertainment, which got the laughs; yet also explored serious themes: what's the right mix of story and science to generate a compelling narrative while still getting the science (mostly) right? Can science onscreen be used to teach science and inspire would-be scientists?

This event had roots in a similar Science and Entertainment Exchange panel at the 2010 American Association for the Advancement of Science (AAAS) meeting. During the audience Q&A, Donna Nelson, a chemistry professor at the University of Oklahoma (currently a visiting professor at MIT), revealed that she advises the AMC show *Breaking Bad.* This is the Emmy-winning saga about a cancer-ridden high-school chemistry teacher who manufactures and sells methamphetamine so he can leave his family in good financial shape. Donna was inspired to organize a panel with a chemistry theme. I helped her connect with The Exchange and her idea got a favorable response from Nancy B. Jackson, ACS president for 2011, who designated it a special Presidential Event and helped plan it.

The resulting panel included Moira Walley-Beckett, writer/ producer for *Breaking Bad*; Kath Lingenfelter, writer/producer for Fox's *House*, whose lead character Dr. Gregory House is a brilliant but controversial medical diagnostician; Jaime Paglia, co-creator/ executive producer for the SyFy channel's *Eureka,* about the sheriff of Eureka, Oregon, and its population of scientific geniuses; Kevin Grazier, who worked on the Cassini mission and consults for *Eureka, Battlestar Galactica,* and other shows; me, Sidney, a physics professor at Emory University who wrote the book *Hollywood Science* about science in the movies and teaches the course *Science in Film*; and Mark Griep, a chemist at the University of Nebraska who teaches chemistry through film clips, as presented in his book *ReAction!*

The panelists illustrated different ways to deal with science on screen. Moira explained that *Breaking Bad* is a drama that involves real science and so must get it right to be convincing—as seconded by Donna, who showed how much research it took for her to back up just one critical line of dialogue with valid chemistry. Kath, a would-be neuroscientist who instead became a science groupie after encountering organic chemistry, similarly discussed the extensive research she does to create the medical mysteries seen on *House*. Both emphasized that correct science or medicine alone is not enough; their writing has to integrate believable emotion, motivation, and action for the characters (as Moira put it, what it takes to turn Mr. Chips into Scarface)—or there's no drama, no story, and no interest.

Jaime, on the other hand, pointed out that *Eureka* is science fiction and so can engage in speculative science like human cloning and time travel. But there is a line, the scientific equivalent of jumping the shark, that the show does not cross. It's the line between a world where things happen because of pure, inexplicable magic, and a world where they happen because of explainable—if only potential—science. Jaime noted that science consultants like Kevin help to keep the two separate. Kevin went on to explain that he's an enabler, not a traffic cop. When asked about a scientific point, he doesn't say "You can't do that!" thus bringing the story to a screeching halt; rather, he says, "That won't work, but this will" and sometimes finds real science that even exceeds what the writers imagined. And whether the science is completely or only mostly right, he added, science fiction has benefits like encouraging innovation.

My talk noted that science fiction films like *Avatar* (2009) are powerful cultural forces that present science with a Hollywood spin. To illustrate, the audience and I had fun with tongue-in-cheek questions like "Where do incoming asteroids always hit?" Hollywood's answer is "Manhattan!" Still, Hollywood science can provide outreach, as when I discussed the science behind the alien bugs in *Starship Troopers* (1997). But maybe someday films will carry a Good Science Seal guaranteeing that "No scientific concepts were harmed in making this film." The last speaker, Mark, also dealt with teaching through film. Movie scenes with chemists and chemistry, he finds, magnetically draw students' attention to the science. He tracked down more than 100 films containing good chemical teaching moments. Many, such as *Apollo 13* (1995), are no surprise, but Mark also unearthed some unlikely gems—the most unexpected, *Clambake* (1967), in which Elvis Presley is a chemist who develops a super hard varnish called GOOP.

All the panelists spoke and answered questions with verve, and the audience clearly enjoyed the entire event. They applauded enthusiastically, a chemistry club from the University of Michigan presented Moira with a T-shirt that said, "We do in class what would be a felony if you did it in your garage," and people flooded the speaker's dais at the end. Like all scientists, chemists love what they do, and the attendees undoubtedly got a lot out of the regular scientific ACS sessions; but I'll bet they had the most fun at *Hollywood Chemistry*, while learning some important lessons about science in entertainment and in society.

Getting the Film Physics Right

Reviews of the films *The Researcher's Article, Conservation, Strange Particles*, *Stuck in the Past, Einstein–Rosen, (a)symmetry,* and *Bien Heureux (All is Well)* (These films can be seen at the Labocine website labocine.com.)

Whenever bad science appears on screen, a physicist is likely to declare "that's wrong!" making it hard to just sit back and enjoy the action along with your popcorn. As a physicist who writes about science on screen, I myself have pointed out Hollywood's errors in physics and other sciences. But now, having reviewed some 150 science-based films, I've learned that the usual rationale for distorting the science is to maintain the flow of the story, which does not automatically make these films scientific disasters. They can still provide vivid teaching moments, publicize real science–society issues, and point young people toward science. Ideally though, the science should receive its proper weight too.

Fortunately, despite Hollywood's tendency to put "story" over "science," there are a number of independent filmmakers who make science an integral part of their stories. Without the publicity and distribution machinery that brings Hollywood features to many millions around the world, however, such independent films typically reach far smaller audiences. But in compensation there are lots of these films, supported by organizations that value their fresh approaches—and some indie efforts become films that are indeed seen by millions.

Physics is well represented among these independent films. With roots in a film festival held by scientist–filmmaker Alexis Gambis in 2006 at The Rockefeller University, his New York-based Imagine Science Films (ISF) is a successful nonprofit devoted to merging science and film. ISF sponsors varied festivals that show independent science-based films around the world and encourages scientist–filmmaker collaborations. In 2016, Gambis began Labocine, an online digital platform with 3000 science-based fiction, documentary, and animated films, accompanied by curated comments.

Many of the films at Labocine.com convey what physics is really like. For example, *The Researcher's Article* (2014) entertainingly shows the process of publishing a physics paper, and how important this is to its authors. *Conservation* (2008) dramatizes what happens when credit for a physics breakthrough is stolen. In *Strange Particles* (2018), a young theoretical physicist, frustrated by his lack of research progress and inability to inspire students, faces a hard question: is there any point in being a scientist if you're not brilliantly talented?

Some films express physics ideas. *Stuck in the Past* (2016) shows how the finite speed of light brings us cosmic history, as an astrophysics student looks down the length of Manhattan and imagines historic moments carried by light that has been traveling since New York City was founded. Touching on general relativity, in *Einstein–Rosen* (2017), two brothers with a soccer ball show that a wormhole allows travel in time as well as space. In *(a)symmetry* (2015), quantum theorist David Bohm talks about the deep meaning of quantum physics; and in *Bien Heureux* (*All is Well*, 2016), a young physicist has no luck in explaining quantum entanglement to a friend, but educates us, the viewers.

The Alfred P Sloan Foundation also supports independent science films. Doron Weber, who directs Sloan's program in Public Understanding of Science, Technology and Economics, sees film as one way to bring science to people. As he describes it, film, together with books, theater, and other media, "support and reinforce each other to showcase stories about science and scientists." The program has provided more than 600 screenwriting and production grants to develop science films and presents awards to outstanding science films. Weber also works with the Sundance Film Institute and other film schools to "influence a generation of aspiring filmmakers to integrate science and technology" into their work by exposing them to science. He finds that most of the 263 Sloan film school awardees continue to work in entertainment media and include science and tech in their creative efforts.

Many of the Sloan-supported films can be viewed online at scienceandfilm.org. Since 2000, about 140 of these have covered physics, astronomy and space science, and mathematics. They include documentaries such as *Particle Fever* (2013), about the first experiments at CERN's Large Hadron Collider, and *Chasing the Moon* (2019), covering the early days of the Space Age.

Biographical films include *Dear Miss Leavitt* (2018), about the pioneering astronomer Henrietta Leavitt, and *Adventures of a*

Mathematician (2019), the story of Polish mathematician Stanislaw Ulam and his contributions to designing the hydrogen bomb and to early computation. Some films with roots in Sloan support have reached millions through wide theatrical release, such as the Oscar-nominated hit *Hidden Figures* (2016), which started as a Sloan book grant.

Figure 4 *Hidden Figures* (2016) tells of the real-life Black female mathematicians who made crucial calculations for NASA in the 1960s. In this scene, one of the group, Katherine G. Johnson (Taraji P. Henson) considers some orbital math while her manager (Kevin Costner, standing center) and others look on.

Asked about the importance of supporting independent films outside the Hollywood mainstream, Gambis and Weber give remarkably similar answers. Gambis notes the varied scientific fields that Labocine films cover and their cultural diversity. For instance, the nine films I described represent five different countries and include four women among their writers and directors. Weber also cites the range of subject matter and genres, and the varied ethnicities and nationalities and high proportion of women among Sloan filmmakers.

For us as physicists, these films have another special value: view one, and you just might learn something new about your science and yourself.

Facing up to Facial Recognition

Reviews of the films *Slaughterbots*, *What Is My Face?*, and *In Vivid Detail* (These films can be seen at the Labocine site http://labocine. com.)

Contemplating your face in the mirror as you brush your teeth in the morning, do you consider its role as you make it through the day? Everyone you encounter, from your spouse to strangers in the street or on Zoom, sees how you present yourself and could make a good guess about your mood. Recognizing faces and the emotions they show is a highly evolved human ability. Computers can also recognize a face, in order to unlock a phone or identify criminals and suspects as seen by surveillance cameras. But algorithmic facial identification by police is now under fire because it is less accurate for Black faces than white ones, causing biased false arrests or worse. Several Labocine films treat facial recognition, both the algorithmic and human kinds.

Slaughterbots (2017) begins with a man dressed like a tech company executive and standing on stage before an audience, but this is no TED talk. He is selling a new law enforcement and military tool, a tiny palm-sized autonomous drone. It is a smart weapon that decides when it has found the right target to kill, using an onboard camera and facial recognition technology to identify a pre-programmed person. The presenter demonstrates how the drone homes in on a dummy standing on stage and delivers death with a small shaped charge that drills right through the temple into the brain. The presentation continues with a video showing real people being executed by the drones; all "bad guys" as the speaker reassures the audience.

These drones can be released in unstoppable swarms. The film moves into pseudo-documentary mode when it shows imaginary TV news about drone swarms killing 11 selected U. S. Senators in the Capitol Building, all from the same party, and killing thousands of university student activists who were targeted through social media. When we return to the presenter, he is ending his demonstration by saying to a wildly cheering audience, "Smart weapons consume data. When you can find your enemy using data...you can target an evil ideology [pointing to his temple] right where it starts." A real-life postscript follows as Stuart Russell, an AI researcher at UC-Berkeley, explains that the film is a warning against the dangers of allowing "machines to choose to kill humans," a warning echoed by many in the AI community.

What Is My Face? (2019) is a welcome counterpoint to what is projected in *Slaughterbots* because it deals with the benign and fascinating topic of human face recognition. It begins with a request: "Please describe your face," followed by the faces of different people who struggle to answer that seemingly simple query.

Two of the faces belong to Sofía Landi, a neuroscientist at The Rockefeller University in New York, and Mark Slutsky, a Montreal-based writer and filmmaker. Together they created this film to show the complexities of faces from different viewpoints. Slutsky's view says much about the human impact of faces: "The essential element of a film is a close-up of a person's face," he says. "It's how you sort of create emotion in movies, visually it's usually through people's faces." Landi studies face recognition, especially "how our brains allow us to recognize people we know... Humans and other primates have areas in their brains that respond selectively to faces." In 2017, Landi reported on facial recognition in macaque monkeys. Using functional magnetic resonance imaging (fMRI), which shows brain activity nearly in real time, she found that familiar faces excite two previously unknown brain areas that extend the brain's known face-processing system.

Landi and Slutsky go on to tell us more about facial recognition, illustrated with images of Landi's research and faces in crowds. Slutsky underlines the power of faces by pointing out that we see them in nonhuman objects like the Man in the Moon, and the remarkable fact that even with radically scrambled features as in a Picasso painting, we know we are seeing a face. Landi explains that it's hard to describe a face because our brains take a holistic view and describes the steps in facial recognition: finding the face in a scene, identifying the person behind the face, and reading the emotions the face expresses. But, she adds, "there is still so much we don't understand." At the film's end, a street artist caricatures Landis and Slutsky together and sums it all up when he says, "A face is your identity and your signature. Everything [is in] your face."

In Vivid Detail (2007) shows what happens when the "everything" a face projects is not received. We first see Justin (John Ventimiglia) on lunch break, trying to assemble a picture puzzle of a smiling emoji face. At the architectural firm where he works, he meets consultant Leslie (Piper Perabo), a woman with an appealing face. After some office interactions, she asks if he wants to discuss his project over lunch, but he begs off. He didn't understand that she was showing interest in him until a co-worker says, "Justin, she asked you to lunch." Later he awkwardly asks Leslie to dinner. It goes well and they share a kiss afterward, but two puzzling incidents occur. He mistakes one blonde waitress for another, and when two of Justin's friends appear, it's obvious to Leslie and to us that they are identical twins, but not to Justin.

The next day, Leslie is hurt when Jason seems to ignore her at work. He finally explains that he sees faces like stick figures that lack unique features. He cannot recognize people when, for instance, they change hair style. He has had this neurological condition, prosopagnosia, ever since he injured a particular part of his brain at age five. It's like being colorblind, he says, where "you can see colors but you can't tell them apart." Leslie seems to accept this, and they are playful about his inability to see faces. But when he has to admit that he can't even tell when she is smiling, they both get angry and frustrated and Leslie leaves.

Later, as Jason walks alone, he sees a street artist carefully sketching a child's face and gets an idea. He asks Leslie to pose for him. She agrees but looks dubious and asks, "What are you doing?" He says, "I want to be able to see you – right now," as he carefully draws her face, detail by detail and square by square, on a cross-hatched surface. And when he asks her to smile, she responds with a gorgeous and radiant smile.

What Is My Face? shows the complexities of facial recognition and that it will take more research to fully understand it. This makes it all the more remarkable that our eyes and brain carry out this intricate task without conscious effort, making visual recognition a vital part of human interaction. *In Vivid Detail* indicates just how vital by showing that those with face blindness or prosopagnosia, about 2% of the population, lose the essential social and emotional benefits of reading faces. Algorithmic facial recognition as seen in the science fiction world of *Slaughterbots,* however, and in our real world of policing, seems to offer more costs than benefits. We need to think less about extending this technology and more about the ethics of using it.

Sunday Is Maroon: Synesthesia on Screen

Reviews of the films *COLORCONDITION, An Eyeful of Sound*, and *Synesthesia* (These films can be seen at the Labocine site https://www.labocine.com.)

One of the many ways that our brains can astonish us is synesthesia, a neurological phenomenon that is well described as "a condition in which ordinary activities trigger extraordinary experiences." I can give a concrete example from my own childhood, when without knowing it, I was a synesthete; that is, my brain automatically generated extraordinary cross-sensory experiences with no effort from me. Whenever I saw or heard the name of any day of the week, I immediately saw in my mind's eye a color, always the same one: Sunday was always deep maroon, Friday green, Saturday a pattern of gray and silver circles, and so on.

A synesthetic interaction can happen between any pair of the five senses. Besides my type, there are synesthetes for whom sound produces colors, words generate tastes, or touch induces smells, and other combinations as well. These strange linkages mean that synesthesia was once misdiagnosed as a delusion, mental illness, or a metaphor stretched too far. It is none of these, but as research has shown, a real, typically stable mental state found in about 4% of the population. For me it seemed perfectly normal.

Synesthesia runs in families, and researchers are tracing its complicated genetic markers. Since the 19th century, it has been associated with creativity and artistic ability. Recent testing confirms that it is more prevalent in people who do creative work. Vladimir Nabokov and Richard Feynman, respectively, saw colors in letters of the alphabet, and in algebraic symbols; David Hockney and Lady Gaga both turn music into colors. But although synesthesia is now better understood, its cause is still incompletely known. It remains semi-mysterious and has always been hard to describe to non-synesthetes. Three Labocine films illustrate what synesthesia is, how it feels, and the surprising connections it generates.

The brief story *COLORCONDITION* (Jason Chew and Rodrigo Valles, 2016) lays out the basics of synesthesia and its human effects. We meet a bearded, casually dressed young man who you might guess is an artist. He speaks to us in a quiet, subdued way: "Some people say what I have is a superpower. Some people think it's strange...I just see things differently. I'm not some kind of monster...I didn't know I had it until I was like 20." The "it" is one form of synesthesia, which he presents by first explaining that the human brain has evolved to interpret certain wavelengths of light as colors. "My brain," he goes on, "also interprets sound as colors. My wires got crossed...my brain [makes] more connections," and adds "Some people compare it to tripping. I guess I'm just tripping all the time."

Colors are important to this young man, who is indeed a painter, but "tripping all the time" affects his whole life. We see one such moment when he both hears a wailing police siren and sees it as "a thousand stars being born in front of your eyes and then dying seconds later." Yet as he paints on a canvas and the woman in his life hugs him, he says "persons are the most colorful and sounds are the most colorful;" perhaps it all somehow comes together in his work and his relationship. The film ends with a beautiful visual summary, a changing pattern of pastel colors with another colored image like a sound spectrum overlaid on it.

The documentary *An Eyeful of Sound* (2010, Samantha Moore) conveys more about the neuroscience of synesthesia, while its colorful animations and voiceovers from synesthetes give an experiential sense of the phenomenon. The film begins with several voices uttering short phrases that appear on a shimmering blue plane of color: "It starts like that...it's almost like a...there's more than one color there...," and others. Then a woman says, "I like muted sounds... softer shoes ...people walking...I feel it in my mouth and I taste it." Now we understand that we are hearing three women find words to describe their synesthetic experiences.

Their continuing descriptions and mantras like "all sounds have color...the alphabet has color" are repeated, juxtaposed, and overlapped to give a sense of varied synesthetic interactions. The accompanying music and animations that are abstract or illustrate synesthetic effects such as dogs barking in color enhance the effect. Some comments show how a synesthete begins to grasp the confusing nature of a mixed response. "I was made much more aware of this," one woman relates, "when I heard an orchestra playing...I thought there was some sort of colored quilt. I couldn't quite make out what was going on." Then attention shifts to psychologist Jamie Ward, a leading synesthesia researcher at the University of Sussex, U. K. His discussion of the science of the effect is interwoven with evocative animations and parallel voiceovers from the women.

In the last minutes of the film, the women explore their experiences among themselves, echoed in imaginative animations of their internal reactions. This builds to a kind of final crescendo of forms and colors in motion. Short of entering the mind of a synesthete, this may be as close as a non-synesthete can get to experiencing the effect.

Despite its name, the surrealistic music video *Synesthesia* (Corey Creasey, Ian Kibbey, and Terri Timely, 2009) isn't about that phenomenon in the same way as the other two films but refers to its sensory experiences and their connections. The film begins with silent overhead shots of fantastic arrays of food sensually arranged in kitchen drawers and heaped on a dining table—sausages, fruits and vegetables, whole fish. As the mother of the family cooks by putting printed words from recipes into the oven, the father, seated in front of a massive stereo wall, eats from a tray that includes a presumably edible book.

Meanwhile, a son plugs his headphones into various foods, which turns on the soundtrack and starts a series of events: balls, vegetables, frogs, cats, vines, and colored shapes emerge from the stereo speakers while they play hard driving music. As these things pile up on the floor, sirens are heard and the speakers emit smoke, and sparks that set the drapes on fire. But the father keeps eating calmly, the mother serves a turkey made of newsprint, and the son plugs his headphones into his navel, then disappears in a puff of smoke as the film ends. Without directly presenting synesthesia, the film's rapid-fire allusions—the sense of taste, as represented by food, turning into sound; the tactility of connecting headphones generating both music and colored objects; words becoming food— resemble synesthetic experiences. Besides, *Synesthesia* is pure fun.

The other two films show the human reactions to synesthesia. Sound–color synesthesia may enhance the artistry of the young man in *COLORCONDITION,* but he has doubts about its overall impact. The participants in *An Eyeful of Sound* often find their unusual internal links wondrous, but also find moments when it would be simpler if these were turned off. My own synesthesia vanished at a young age, but in retrospect it seems purely enriching. Maybe if I had lived with it or another type for many years, I too would find my reactions to it complex, like synesthesia itself.

Science Cinema Online: The 13th Annual Imagine Science Film Festival

Reviews of the films *Coded Bias*, *The Last Artifact*, *FREYA*, *Pripyat Piano*, *Story*, *Spanish Flu: The Forgotten Fallen*, *The Atomic Adventure*, *La Bobine 11004*, and *Coronation* (These films can be seen at the Labocine website labocine.com.)

When Alexis Gambis founded Imagine Science Films (ISF) over a decade ago, he could not have foreseen that its 13th annual film festival would occur during a pandemic that underlines exactly what ISF wants to do: that is, make people aware of what science means in our lives, an understanding heightened today as we face a deadly coronavirus that only science can stop. Fortunately, like other institutions, ISF found a way to safely and effectively present Imagine Science Film Festival 13. This event was shown fully online using the Labocine platform, a library of science-based films, between October 16 and 23, 2020.

From Hollywood mega-studios to independent films and film festivals, the film world is grappling with the loss of the big screen, live audience theatrical experience during the pandemic (and maybe after). ISFF13 showed that there are some plusses to going online, at least for a festival. The online version presented 80 films, several times the number displayed in earlier years, with many enriched by remote follow-up panel discussions with the directors and others. The films had over 20,000 total views, a nine-fold increase over the 2019 in-person festival.

The online effort worked smoothly, and its films reflected what ISF has supported over its history. They represented fiction, non-fiction documentary and experimental, and animated films. They displayed wide diversity, coming from 41 different countries with their varied ethnic groups. The film's directors were nearly equally balanced between men and women, with more women at 54% (while Hollywood continues to struggle with gender equality). There was scientific diversity too. The films covered topics from animal behavior to general relativity, genetics to botany and related to the pandemic as well in *Spanish Flu*, about the great pandemic of 1918, and *Coronation*, about the response to COVID-19 in Wuhan, China.

And while science was a main topic, so were scientists, with films that showed their humanity and how they think.

Few viewers would have been able to watch all the films, but any viewer could pick a satisfying menu from among the offerings. My own menu leaned toward my scientific research area, physics, and other topics in science and technology that I have written about. Another criterion was to watch films that tackle big issues about how science and society affect each other.

That last criterion applies to the feature-length documentary *Picture A Scientist* (2020, USA, Ian Cheney and Sharon Shattuck, 97 minutes) that opened ISFF13 and was a selection of the 2020 Tribeca Film Festival (canceled, however, due to the pandemic). The title is a reminder of the Draw-A-Scientist test where young students are asked to draw a scientist, as a way to explore early perceptions of researchers. Typically, both boys and girls draw a white male in a lab coat, often with unruly hair like Einstein's. Challenging this stereotype animates the three PhD female researchers in the film who describe the personal and professional barriers in their careers. Their strong narratives mingle with stories and analysis from other women and some men, filling in the picture of how badly women have been treated in science and how to change that.

Nancy Hopkins, professor of biology at MIT, relates her harassment by a Nobel Laureate scientist, then how her awareness of unfairly allocated research resources led her with others to improve the status of women at MIT. Raychelle Burks, then professor of chemistry at St. Edward's University, Austin, Texas and a woman of color, speaks of harmful moments in a profession dominated by white men, from being mistaken for a custodian to discomfort at scientific conferences. She embraces the value of diversity in science by being authentically herself in award-winning outreach, to demonstrate that scientists are not only white men. Jane Willenbring, professor of geology, Scripps Institution of Oceanography, tells of humiliating sexual harassment by her adviser David Marchant during field research in Antarctica. Her formal harassment claim years later was triggered by the fear that her young daughter could one day be "treated like trash" as she had been and led to Marchant's firing from Boston University.

Another feature-length documentary, *Coded Bias* (2020, USA, Shalini Kantayya, 86 minutes) explores a growing issue in our society: are decisions about people made by algorithms and AI correct, fair, and equitable? It might seem that logical and intelligent algorithms are more objective than potentially biased humans, but according to *Coded Bias*, this is a poor assumption. As the film opens, Joy Buolamwini, a computer researcher at the MIT Media Lab, discovers that standard facial recognition software, as used by police and in other applications, does not properly recognize her own dark-skinned face.

Figure 5 The documentary film *Coded Bias* (2020) explores the dangers of facial recognition and surveillance technology, especially for people of color. Here computer scientist and activist Joy Buolamwini finds that facial recognition software does not detect her dark face until she dons a white mask.

The film shows other examples of algorithms that fail to make correct or meaningful judgments, or undermine privacy, civil rights, and racial equity: improperly rating an experienced, award-winning teacher in Houston; racially profiling ordinary citizens for police scrutiny in the U. K.; and enabling the Chinese government to closely scrutinize and evaluate its citizens in their daily lives. *Coded Bias* was in production before Robert Williams, an African-American man, was unjustifiably arrested and jailed in Detroit because of an incorrect identification by facial recognition software. This case perfectly illustrates the dangers *Coded Bias* points out, but the film ends on a hopeful note as Buolamwini and other activists testify before the U. S. Congress, seeking legislation to control the use of algorithms.

A third long documentary, *The Last Artifact* (2019, USA, Jaime Jacobsen and Ed Watkins, 56 minutes), describes a less dramatic but significant interaction between society and science, specifically metrology, the science of measurement. This underlies much of how we and society function, from measuring the area of a rug to setting manufacturing standards and underpinning quantitative science. The film explains how all measurements trace back to universal standards for the physical fundamentals of length, time, and mass. The first two are now defined through the speed of light, and the frequency of the light emitted by a cesium atom. These are constants of nature embedded in the fabric of the universe: But the standard kilogram mass had remained as "the last artifact," a human-made hunk of platinum–iridium alloy kept in Paris.

Then in 2011, scientists began to consider relating the standard mass instead to another fundamental quantity, the constant denoted by h that Max Planck derived in 1900 as the basic unit of quantum mechanics. This would link all physical units to inherent properties of nature, a dream of Planck's as well. The film shows the ultra-precise measurements that determined h to within 13 parts per billion, ending with the vote in 2019 by an international scientific commission to adopt this final part of a new standard system of units. As one metrologist says in the film, this was no less an accomplishment than taking on the task of defining objective reality.

Many shorter entries at ISFF13 make their own important points, especially when viewed in thematic groupings. For example, 75 years after the atomic bombings of Hiroshima and Nagasaki in World War II, and 34 years after the Chernobyl nuclear catastrophe, three short films show in varied ways how the terrors of unleashed nuclear energy still linger.

The documentary *La Bobine 11004* (Reel 11004, 2019, France, Mirabelle Fréville, 19 minutes) traces back directly to Hiroshima. After Japan surrendered, the U. S. government sent in teams to film the effects of the bomb, including a "reel 11004" shot in Hiroshima on April 5, 1946. According to director Fréville, this was kept secret until she found and edited it to make *La Bobine 11004*. Several sources such as the recent book *The Atomic Doctors* confirm that the U. S. worked to suppress evidence that the bomb's radiation greatly harmed Japanese civilians. The original newsreel-like footage in *La Bobine 11004* shows many Japanese children and adults being treated for burn-like lesions and other injuries. Most memorably, it shows a woman with disfigured arms and face looking directly and stoically into the camera.

The Atomic Adventure (2019, France, Loic Barché, 26 minutes) fictionalizes France's 1961 test of its fourth atomic bomb. We follow seven soldiers sent to measure radioactivity after the detonation in the Algerian desert. Unaware of the risks, they josh each other on the way to the site, but their Captain, an older war veteran, is wary. The film dramatizes actual blunders in France's atomic testing as it shows the soldiers, wearing flimsily inadequate protective gear, become exposed to fallout when the wind shifts from its predicted direction. To protect them, the Captain buries them under the desert sand but sees that the radiation has already doomed him. Perhaps also not wanting to participate in this new nuclear age, he walks into the falling radioactive dust. One of the soldiers also dies, underscoring the tragedy of nuclear war at every level.

A more recent nuclear catastrophe inspired the documentary *Pripyat Piano* (2020, Czech Republic, Eliška Cílková, 18 minutes). The 1986 Chernobyl reactor accident in what is now Ukraine widely spread radioactive material and forced the evacuation of 50,000 people from the city of Pripyat. The film memorializes their lost homes and way of life by showing the deteriorating pianos left behind. Mingled with scenes of emergency crews frantically coping with the disaster, we see abandoned upright pianos leaning against walls, on their backs and with exposed insides, and a grand piano still gracing a wrecked auditorium. Some pianos produce only thuds, but others are somewhat playable. Their musical sounds, and background songs from former Pripyat citizens, speak to resilience in the human spirit and human culture. Resilient too is nature itself, as exterior shots show lush forest greenery slowly reclaiming the abandoned city.

Two other short films examine our dependence on and our inability to escape the world of digital devices and of information steadily flowing in, but also being taken from and about us. *Story* (2019, Poland, Jolanta Bańkowska, 5 minutes) comments on our digital lives through amusing animation. The film's hero wakes up, smartphone in hand, to a day full of the digital and the virtual. We see a mother and her child on a swing each pull out a smartphone; a restaurant where each diner is typing on a laptop; and a man play an invisible virtual reality piano, then bow to his online audience. VR also enables a man to play virtual fetch, until his angry dog leaps for his throat. In the funniest and most telling scene, several men in a commuter train each raptly stares at his phone. Suddenly the head of one of them bursts into flame, but the only response is that one of the others raises his own phone to take a video. Does *Story* remind you of anyone you know?

The short drama *FREYA* (2020, Canada, Camille Hollet-French, 17 minutes) richly imagines a near future dominated by digital control of people. As 30-something Jade (Rhona Rees) returns home, she is greeted by the pleasant female voice of FREYA (Federally Regulated Enquiry and Yield Assistant). FREYA is a ubiquitous and powerful government AI that constantly gathers information ("rate your work day") and gives directives ("don't drink too much wine"). Sex too has been digitized beyond today's Tinder, as Jade chooses a lover from the menu of men and their anatomies in the Nookie Bookie app. She enjoys her one-night stand but is dismayed when FREYA, tracking her physiology, announces that she is pregnant and will be under added scrutiny in a society that bans abortion. Jade, however, miscarries, freeing her from unwanted motherhood and inspection but leaving her emotionally drained and realizing that there is no escaping FREYA.

The eight films I've covered give only a partial cross section of what ISFF13 offered. The remaining films are well worth seeing, however, possible on Labocine or elsewhere, and maybe not too long from now, again in the company of fellow fans of science cinema.

Aliens: Love Them, Hate Them, or Relate to Them?

Review of the film *District 9* (2009) and others.

What with black holes, dark energy, and so on, it's a big, strange universe out there. Science fiction films add more strangeness when they include weird and wonderful aliens. The closest we've come to real aliens so far is evidence of water that could support life, past or present, at a few sites in our solar system. But we have recently found lots of extrasolar planets, and that helps fuel a long tradition of speculating about life in the universe.

Moon creatures appeared in the first science fiction flick ever, the French *Voyage dans la Lune* (*Voyage to the Moon*, 1902). In 1924, the Soviet film *Aelita: Queen of Mars* showed an intricate Martian society. Since the 1950s, other new and improved models of aliens have appeared in films—literally improved, since the steady development of special effects, especially computer-generated imagery, has made even extremely strange aliens look believable on screen.

Plenty has been written about what our fascination with aliens says about ourselves. Some aliens are human-seeming (like the alien emissary in the 1951 and 2008 versions of *The Day the Earth Stood Still*), but they're mostly an unsettling lot, which may reflect human paranoia. It's hard to warm up to the slavering reptilian predator of *Alien* (1979), the insectoid creatures of *Starship Troopers* (1997), or the monkey-like bestial aliens of *War of the Worlds* (2005). With their hostility to humanity, these ugly customers return our repulsion in spades. That's not to ignore the few appealing aliens that have appeared on screen. Movie viewers found the lost alien in *E. T.* (1982) endearing despite his funny looks, and the Disney-like Ewoks in *Star Wars Episode VI: Return of the Jedi* (1983) are cute and fuzzy.

Things get more interesting though when we're asked neither to exclaim "how cute!" or to blow away an alien, but to actually relate to one, as in the just released film *District 9*. The film imagines a world in which over a million alien beings have been sequestered for years in District 9 outside Johannesburg, South Africa. This has plenty of parallels with South Africa's racial history, complete with shanty towns, gang warfare, and military, legal, and quasi-legal maneuvering to control the aliens.

But these tall insect-like bipeds with tentacles around their mouths aren't just a little bit different; they're truly inhuman and truly unappealing. Yet they have intelligence, parental feelings, a sense of justice and of loyalty, and a desire to leave their squalid conditions and return home. We discover this as the human in charge of resettling District 9 finds himself more connected to the aliens than he ever wanted to be. He works together with one of them for common goals, and though they don't exactly become friends, they bond.

This can be read as a parable about black–white interaction or more broadly about how to accept what is unfamiliar to us. Either way, the "alien other" carries messages we need to hear. That's one reason why I hope that if and when we find alien life, it will be more advanced than a smear of lichen on a Martian rock.

Trapped on Mars

Reviews of the films *Aelita: Queen of Mars, Robinson Crusoe on Mars, Stranded, The Martian* and others.

Our planetary neighbor Mars has long been one of humanity's favorite celestial bodies. Its red tint made it stand out among the five planets ancient people could see by eyeball alone. When later it was examined by telescope, its polar ice caps suggested that water and maybe life existed there; and in the late 19th century, the American astronomer Percival Lowell thought he discerned huge canals on Mars, presumably built by intelligent beings to distribute water.

Lowell, it turns out, was sadly mistaken, but Mars still fascinates us. NASA has sent unmanned spacecraft and robotic surface rovers to explore the planet and seek signs of life since 1964 and is now considering a manned Mars mission. Science fiction has kept pace with the science, from H. G. Wells' *The War of the Worlds* (1895), about Martians invading Earth, right up to the hit 2015 film *The Martian* about a manned expedition to Mars, based on a book by Andy Weir and nominated for seven Academy Awards, including Best Picture and Best Actor.

The Martian is one of some three dozen films and TV series or episodes that have been set on Mars, beginning in 1910 with the four-minute movie *A Trip to Mars* from the Thomas Edison film company. In some of these, human explorers encounter Martian aliens; for instance, in the film *Aelita: Queen of Mars* (1924), where space travelers from the Soviet Union interact with human-like Martians, and *The Angry Red Planet* (1959), where astronauts find carnivorous plants and other Martian monsters. Later, as scientific findings confirmed that Mars was not a promising abode for life, *Total Recall* (1990) with Arnold Schwarzenegger and *Red Planet* (2000) with Val Kilmer instead portrayed Mars as colonized by humanity.

Some films took a different tack in a subgenre we might call "marooned on Mars." In these, a Mars expedition runs into trouble that leaves one or more of its members stranded on the planet, desperately trying to survive until a rescue spacecraft can arrive. That's the theme of *The Martian*, where "the Martian" is NASA astronaut and botanist Mark Watney (Matt Damon). He is left behind on Mars after a violent storm threatens the spacecraft meant to return him and his colleagues to Earth. Because of a technology malfunction, the others think he is dead and blast off to save themselves. But Watney is alive though injured, faced with surviving in a hostile environment that lacks air, water, and food. This is realistic and exciting science fiction, but it is not the first film about human survival on Mars.

(a)   (b)

Figures 6   Stories and films about the planet Mars have a long history. (a) In the silent film *Aelita: Queen of Mars* (1924), Soviet travelers to Mars find a human-like society where workers are mistreated. Helped by Queen Aelita (Yuliya Solntseva), the visitors and workers overthrow the Martian rulers. (b) Nine decades later, the red planet features in *The Martian* (2015), where abandoned and space-suited astronaut Mark Watney (Matt Damon) struggles to survive alone in its bleak environment.

In 1964, 5 years before humanity set foot on the Moon, the feature film *Robinson Crusoe on Mars* updated Daniel Defoe's classic story of a castaway on a desert island through Mars astronaut Kit Draper (Paul Mantee). As he and a co-astronaut orbit Mars in their Gravity Probe 1 spacecraft, they narrowly miss a piece of space debris and must abandon ship in individual capsules. Draper's fellow astronaut crashes and dies, but Draper lands safely on Mars, leaving him alone except for Mona, a small female monkey brought along as a test animal. Both must struggle to remain alive until rescue can come.

Later, in the Spanish-made film *Stranded* (released in English in 2001, and in Spanish as *Náufragos* in 2002), as the first manned mission to Mars descends to the surface from their orbiting mothership *Ares*, their lander crashes due to an altimeter error (something similar really happened in 1999, when NASA's unmanned Mars Climate Orbiter was lost because of a mix-up in measuring its altitude in units of miles versus kilometers). One crew member is killed, and the pilot of the *Ares* has no choice but to return to Earth, leaving five survivors in the toppled lander.

When they calculate that their life support can maintain only two people during the months until a rescue ship can arrive, the lander pilot Susana Sánchez (Maria Lidón, who also directed the film) and two others walk off into the Martian desert. (This echoes a famous tragedy in the Scott Antarctic expedition of 1912, when Captain Lawrence Oates, fearing that his injuries were keeping his companions from reaching a critical food cache, sacrificed himself by leaving their tent. The others died anyway soon after.)

The expeditions in *Robinson Crusoe on Mars*, *Stranded*, and *The Martian* and their marooned astronauts all have the same needs, beginning with a way to reach Mars and return safely to Earth. Then they must endure existence in a desolate environment where temperatures average –62 degree Celsius, where there is no oxygen to breath (the thin Martian atmosphere, mostly carbon dioxide CO_2, could not sustain humans) or water and food.

Despite these issues, one seeming advantage is Mars' low gravity, only 38% of the Earth's. That would reduce the work of establishing a base and surviving, but it has a downside. Long-term exposure to low gravity, as happens to the crew on the International Space Station orbiting the Earth, has bad bodily effects because joints and muscles weaken when they do not work against the Earth's gravity. The "marooned on Mars" films do not simulate the low gravity on Mars, probably for the practical reason that this would increase the cost and complexity of making the film, but it is one more potential Martian hazard.

The films clearly display the other difficulties of reaching Mars and surviving on it. A manned mission there would be much harder to carry out than our furthest manned space venture to date, which put people on the Moon in 1969. The Moon is 384,000 kilometers away, but Mars is over 140 times further even at its closest approach to Earth, 55 million kilometers. This means eight months of travel one way over the most favorable route. The astronauts would then have to remain on Mars for 10 months until the two planets have again moved into the best position for a return trip, for a total expedition time of over 2 years, 26 months.

That wasn't the plan in *Robinson Crusoe on Mars*, where the astronauts observing Mars from space are forced to land only when their near collision disrupts their orbit. The film's advertising proclaimed "This film is scientifically accurate. It is only one step ahead of present reality!" The movie really was ahead of its time in audaciously sending people to Mars even before we had reached the Moon; but it was behind its time in sending a whole manned mission just to survey the planet. That same year, NASA launched its unmanned Mariner 4 spacecraft to examine Mars without risking a human crew and using cheaper and simpler technology.

The film's accuracy also suffers because the Mars that Kit Draper encounters doesn't much resemble the Mars we now know. The methods Draper uses to deal with low temperatures and a diminishing oxygen supply are simply unrealistic. He shelters in a cave where, by pure luck, he finds yellow rocks that burn like coal to release heat and oxygen that keep him alive. He notices when the monkey Mona apparently discovers a water source and follows her, to find an open pool of water with edible plants growing in it. It may have seemed in 1964 that Mars could provide for Draper's needs; but only a year later, Mariner 4 was sending views of the planet that showed a dead and barren world without water or plant life.

Stranded and *The Martian* display this more realistic assessment of Mars while also postulating missions that go beyond what NASA can deliver today. Besides sending a crew on a long space voyage, NASA would have to send tonnes of supplies for their stay on Mars—a habitation, food, water, oxygen, scientific equipment, and rocket fuel for the return home. So NASA is working on rockets powerful enough to haul all this off the Earth. One scenario is to send the cargo in unmanned spacecraft before the crew takes off, pre-position it on one of Mars' tiny moons Phobos and Deimos, and then land everything on Mars after the crew arrives.

To reduce the cargo load, NASA is also working to make Mars missions self-sufficient by using Mars' own resources, an approach that appears in *Stranded* and *The Martian*. We have found that Mars has some water, oxygen, and methane in its atmosphere and soil. With the proper chemical approach, water and oxygen can be extracted for life support, and methane can be extracted for use in rocket fuel. Knowing this, the scientists in *Stranded* try but fail to build a chemical reactor to make water. Nor can they extend their life support by recycling air and water as originally planned because of problems with their thermoelectric power generator, a real NASA device that converts heat generated by nuclear decay into electricity. These are the reasons that only two astronauts can be maintained in the wrecked lander.

But in *The Martian*, Mark Watney uses similar methods to literally live off the land. He has equipment to make oxygen from the atmosphere, and he has some water. He does not have enough food to last until he can be rescued, but he knows how to grow more even on Mars. He plants potatoes in 126 square meters of sterile Martian soil inside his habitat, which he fertilizes with his own personal organic material, his feces. To get water for the potatoes, he combines oxygen with hydrogen, which he obtains by burning the remaining gallons of rocket fuel hydrazine. This is enough to raise crops of potatoes, which with his stored food gives him 1500 calories a day, the minimum needed to maintain health and life.

Stranded and *The Martian* attack the problem of survival on Mars with valid science that reflects what we have learned about Mars and about space technology in the last half-century. Both films combine an imagined manned trip to Mars with the best available science to give a meaningful image of the planet. But our knowledge of Mars is growing so fast that even *The Martian* is already dated. In the same week it was released, NASA announced it had for the first time found liquid water on the surface of Mars. If and when we send a real mission to Mars, its astronauts may be able to directly obtain drinking water, increasing their odds of success.

Those future NASA astronauts might be encouraged also by the mostly happy endings for stranded movie astronauts. In *Robinson Crusoe on Mars*, Draper encounters an alien but human-like slave who he liberates from his masters. He calls his new friend Friday, like Robin Crusoe's faithful companion in Daniel Defoe's story. They help each other survive until a ship arrives from Earth to rescue them. In *The Martian*, Watney also survives and safely returns to Earth aboard a NASA rescue craft.

But in *Stranded,* only three of the five stranded astronauts make it. The three who left the lander reach the rim of Valles Marineris, a real and spectacular canyon kilometers deep. One astronaut runs out of oxygen and dies as they climb to the bottom, but the others find water vapor and an artificial structure containing air and mummified aliens. A last accident kills a second astronaut, but Sánchez survives. She radios the two left at the lander to join her, realizing that there is enough life support in the canyon to keep them all alive until a rescue ship can arrive.

This science fiction ending is unsupported by the known science of Mars, but it reminds us that though we know the Red Planet better than ancient people did, we still do not know it well. For all of NASA's highly developed technology, for all its careful planning, Mars is a distant and alien place with dangers and mysteries we may not even be able to imagine. And that, of course, is why we want to go there.

Science Advances and Science Fiction Keeps Up

Reviews of the films *The Day After Tomorrow, Geostorm, Blade Runner,* and *Blade Runner 2049.*

When science fiction is done well, it plausibly projects current science into the future, showing where new science and the technology it inspires might take humanity. Two Hollywood science fiction films this year, *Blade Runner 2049* and *Geostorm,* elicit this kind of speculative thinking. *Blade Runner 2049* has received ecstatic reviews, whereas *Geostorm* has been rated only as a run-of-the-mill example of the global catastrophe genre. Nevertheless, both stand out because each can be linked to an earlier film, showing how science fiction follows science to illuminate where we are going.

Blade Runner 2049 is a direct descendant of *Blade Runner* (1982), whose reputation has grown since its release until it has become a classic. Its visual power comes from its memorably moody portrayal of a future Los Angeles in 2019 as a mixture that film critic A. O. Scott calls "neon-*noir*," with giant advertising displays overlaid on a dark, perpetually rainy urban underworld. The emotional and intellectual power of the story comes from its premise that it is possible to create "replicants," synthetic humans barely distinguishable from real ones.

These are manufactured by the Tyrell Corporation to function as slave labor on distant planets. As a safety factor, they are designed to automatically terminate after 4 years. Aware of this limit, a group of rogue replicants and their highly intelligent, exceptionally strong leader Roy Batty (played by Rutger Hauer) illegally return to Earth to try to get their lives extended. Rick Deckard (Harrison Ford) is a "blade runner," a special agent assigned to hunt them down. Deckard terminates the others, then watches Batty reach the end of his 4-year span and die in a famous scene that shows how the replicants reach for and nearly achieve true humanity.

While hunting the rogues, Deckard meets Rachael (Sean Young), a beautiful, advanced model female replicant. She has had artificial memories implanted, giving her a seemingly real personal history that makes her believe she is human. Deckard and Rachael connect romantically, and in both the film's original theatrical release and its later re-edits, they go off together at the end of the story. This ending and other hints in the film have led to much speculation that Deckard himself is a replicant. The ambiguity has never been resolved and adds weight to the film's big question, "What does it mean to be human?"

The other unresolved question, "What happened to Deckard and Rachael?" becomes important in *Blade Runner 2049*, set 30 years after the original film. The Tyrell Corporation is gone, and replicants are made by industrialist Niander Wallace (Jared Leto). He believes these slaves are essential for humanity to spread into space but cannot make enough of them to meet the need. However, their technology has improved and advanced replicants are now used to hunt down rogue replicants. Replicant Agent K (Ryan Gosling) is one of these new blade runners, working for the Los Angeles Police Department. Adding another element to the mix of real and synthetic humans, K's live-in girlfriend Joi (Ana de Armas) is a holographic projection with artificial intelligence (AI) who seems to share real and mutual feelings with K.

Figure 7 *Blade Runner 2049* (2017) takes replicants, the artificial people in the classic original *Blade Runner* (1982), several steps further. Agent K (Ryan Gosling) is a replicant who hunts down rogue replicants for the LAPD and finds that they can now reproduce biologically. In this scene, he talks to another kind of artificial person, his holographic girlfriend Joi (Ana de Armas).

The story begins when K terminates a rogue replicant, then searches the rogue's property and finds a buried chest. It contains an apparently human skeleton from a woman who has given birth by emergency Caesarean section, but closer examination reveals a serial number etched on a bone, showing that the skeleton came from a replicant. This is a bombshell revelation because replicants are supposedly unable to reproduce. If they can, they are that much more human and perhaps even possess a soul (whatever that may mean) with destabilizing effects on human society. To prevent this, Agent K's commander at the LAPD orders him to destroy the evidence and trace the replicant child. K finds himself in a race with Niander Wallace who also seeks the child and the secret of replicant reproduction so he can create a self-propagating slave race.

K determines that the skeleton is Rachael's, and learning about her long-ago relationship with Deckard, K tracks him down in the abandoned city of Las Vegas. Deckard (Harrison Ford again) confirms that Rachael became pregnant, but he has never seen the child, who went into hiding. Meanwhile, Wallace's agents arrive, shoot and wound K and bring Deckard to Wallace who is unsuccessful at learning about replicant reproduction from him. K rescues Deckard and, in a moment of clarity, realizes that an expert he had consulted about implanted memories is Deckard's grown-up missing child. As the film ends, Agent K takes Deckard to meet his daughter, then lies back, and expires from his wounds.

When *Blade Runner* was made in 1982, robotic or biological technology that could produce synthetic humans was comparatively primitive. Researchers had started constructing humanoid robots in the 1970s. These have now reached the point that to some extent they resemble and move like real people and carry out a degree of cognition, but they remain too limited to be taken for human. Recent advances in AI, computer vision, simulated skin, and so on show that we may someday be able to create truly life-like humanoid robots, if we should ever want to; but today's robotic technology, let alone that of 1982, cannot accomplish this. However, AI applications like Apple's SIRI and Amazon's Alexis, digital assistants that take in and respond to human speech, have put us on the road to making non-physical companions like Agent K's girlfriend Joi. This was not even a remote possibility in 1982.

Also in 1982, the structure of DNA was known but the human genome had yet to be mapped. Once that was done in 2000, possibilities exploded for genetic manipulation of human DNA and even for synthetic biology—the design and construction of artificial cells, tissues, and whole organisms. The CRISPR technique, discovered in 2015, makes it easy to modify human DNA. In 2017, this was used in apparently successful trials to eliminate a genetic mutation that would cause sudden cardiac arrest and death. Many scientific and ethical issues must be resolved before genetic engineering can be widely used, but its potential benefits are bringing major scientific interest and business investment. It will surely continue to develop rapidly.

If one wants to create a race of people with specified characteristics, it would be easier to modify existing human DNA then to build whole synthetic humans, humanoid robots, from the ground up—especially if the males and females are meant to couple and create children, which seems far outside what robotic technology could do even in imagination. If the replicants in the 1982 *Blade Runner* were not robots but products of modified human DNA, that would explain how Deckard and Rachael produced a child, no matter whether Deckard is a human and Rachael is a replicant or they are both replicants. In fact, *Blade Runner 2049* shows that Rachael was essentially human, not a robot, because her skeleton is an ordinary human one—and why would it be useful or necessary to reproduce every internal human feature in a constructed robot?

The implicit message of *Blade Runner 2049* is that the biotechnology to produce replicants has existed since the time of the Tyrell Corporation. Niander Wallace's dream is to capture that technology to create a race of slaves whose DNA has been modified to make them strong and docile workers. This kind of thinking is like that from the discredited notions of eugenics, Nazi "racial purity," and today's white nationalist movements. *Blade Runner 2049* projects not only current science into the future, but elements of our current society.

Another scientific progression appears in two films about climate change: *Geostorm* and before it in 2004, *The Day After Tomorrow* (*TDAT*), which seamlessly segues into *Geostorm* like a prequel. In *TDAT*, climatologist Jack Hall (Dennis Quaid) uses data obtained in Antarctica to conclude that global warming will soon trigger a new ice age (this seems contradictory but is a real possibility). Despite strange weather such as snow falling in India, the U. S. government rejects Hall's warning about the coming catastrophe. But he is right, and as planet-wide storms produce tornadoes in Los Angeles and a tsunami that inundates Manhattan, the temperature drops and ice covers much of the Earth. While U. S. inhabitants belatedly move south to ice-free Mexico, Hall treks from Washington, DC to Manhattan where his son and others are trapped by ice. He saves them, the global storm passes and the ice melts, and the world returns to normal.

Geostorm could take place a few years later. It begins as if in the interim the world had not yet dealt fully with climate change and extreme weather—but now, rather than respond passively, humanity finally decides to control these effects through geoengineering. The nations of the Earth cooperate to build a global network of space satellites that use heat, pressure, water, and sonic waves to manage weather. The system works well until suddenly it doesn't as even more destructive phenomena appear—tsunamis and floods, drought, instant freezing of a village and its inhabitants in the middle of a desert, and ferocious lightning. The weather network, nicknamed "Dutch Boy," has gone out of control and is creating a worldwide geostorm.

Dutch Boy's chief architect Jake Lawson (Gerard Butler) and his brother Max (Jim Sturgess) find that the system has been sabotaged by a rogue group in the U. S. government that wants the United States to emerge as the world's absolutely dominant nation, just as it was in 1945. The brothers find the leader of this group and turn him over to the authorities. Jake resets Dutch Boy to function properly, and the world is again at peace.

In the 13 years since *TDAT* was released, the power of extreme weather events and other effects of climate change has increased. Temperatures have risen, storms like the recent hurricanes are more intense, and some coastal areas are regularly flooded by sea level rise. New York City is projected to experience what were once-in-500-years floods every 5 years by 2030. The U. S. Government Accountability Office recently estimated that extreme weather and fire events, amplified by climate change, have cost the federal government $350 billion in disaster relief over the past decade.

Geostorm is certainly timely, and the film goes beyond *TDAT* by introducing geoengineering to counter climate change, though so far, real geoengineering is not the use of space satellites to control weather. Rather it is the use of planet-wide methods to remove the atmospheric greenhouse gas carbon dioxide, for instance by enhancing the growth of microscopic plants (phytoplankton) in the ocean that take in carbon dioxide; or to decrease warming by making the Earth absorb less solar radiation, for instance by putting aerosols into the atmosphere that reflect sunlight back out into space.

These approaches would be less effective than reducing human emissions of greenhouse gases, but geoengineering could be one of several important tools. However, the approach is controversial because it is hard to predict all its global consequences, and once negative outcomes start, they may be impossible to stop. There is a political aspect as well. Advocates of a global response to climate change fear that emphasis on geoengineering will serve as an excuse to ignore action to reduce greenhouse gas emissions like the Paris Agreement on climate change.

Geostorm carries a scientific warning, which is that any form of global control of climate, whether by space satellites, ocean biology, or aerosols, must be carefully considered before it is implemented; and also a hopeful message, that at least in fiction the world's nations can decisively act together to face a global threat. *Blade Runner 2049* continues the scientific, philosophical, and moral debate about what it means to be human. It reminds us too that the technology to modify and create new forms of humanity is becoming a reality, whose consequences for both the old and the new humans we are barely beginning to understand.

Entropy and the End of the World in *Tenet*

Movies have always played with time. The director tells you where to look and how time behaves as you watch, starting with dividing the story into scenes, then choosing how to move between them, from slow fades to fast jump cuts that set different rhythms. Screen time can be slowed, quickened, or reversed and studded with flashbacks and flashforwards. This nonlinearity makes film the ideal medium to tell stories about physically altering the stream or direction of time, or how we perceive it. Nearly 240 films have done either in the last century but the latest, *Tenet* (2020), may have twisted screen time to the ultimate, pleasing some critics and viewers and baffling or annoying others. For me, the film's story didn't jell, although it did provoke thoughts about science, film, and time.

Tenet follows themes in director/writer Christopher Nolan's earlier work. *Memento* (2000) is a psychological thriller about a man who misperceives his personal time through recurring short-term memory loss, which he replenishes with photos to track what he can't remember. The film represents this condition with black-and-white sequences in chronological order, and color sequences in reverse order. *Interstellar* (2014) is a science fiction film about humanity traveling through a wormhole to find new planets. Besides compressing distances, wormholes that connect black holes also skew time. That becomes an emotional story element as the time distortions allow a lead character to appear as a "ghost" to his young daughter in a different temporal era. The film's science credentials were burnished when Caltech theoretical physicist Kip Thorne consulted for it and wrote a related book.

IMDB calls *Tenet* a science fiction action-thriller, although it is less sciency than *Interstellar*, and the blinding level of action overwhelms any personal stories it might have developed as in *Memento*. For *Tenet*, Kip Thorne gave only limited help. "I promised him," said Nolan, "I wasn't going to bandy his name around as if there was some kind of scientific reality to *Tenet*. It's a very different kettle of fish to *Interstellar*." That's an apt comment, yet Nolan did bring in one science concept about manipulating time that I've never seen before in a film: entropy. The film also mentions some of the human, if not personal, philosophical and ethical conundrums that bending time would create.

It isn't easy to summarize *Tenet* since its convoluted story slowly reveals itself over two-and-a-half hours. But writer Nate Jones has heroically produced a 3400-word nearly frame-by-frame recap of the film [2]. Supplementing my own viewing with his invaluable guide, here's the CliffsNotes version.

The film begins as armed terrorists invade the opera house in Kiev, Ukraine, and capture a CIA agent played by John David Washington (who is only ever given the name "Protagonist") after one inexplicable event: an unknown, black-clad figure kills Protagonist's attacker with a bullet that seems, well, to travel backward! The captured Protagonist resists torture and swallows a suicide pill but doesn't die. He awakens as his CIA manager tells him that the attack was a sham designed to make Protagonist vanish, and that having passed the loyalty test of taking the pill, he is just the man to join Tenet. He learns more about what he's joining when he visits a white-coated CIA scientist named Barbara played by Clémence Poésy.

She explains that the strange bullet was one of many "inverted" objects that travel backward through time. Protagonist is convinced when he watches an inverted bullet he had dropped fly back up into his hand without his touching it. He asks how effect can precede cause, and what this says about free will. Barbara blithely assures him "no problem," because whether the event runs forward or backward, he caused it. Two important facts emerge: inverted objects travel back through time because their entropy runs backward, as Barbara explains with a bit of movie science-chat: "We think it's a type of inverse radiation triggered by nuclear fission;" and these objects indicate that a terrible, totally destructive war is coming from the future.

Protagonist and his newly acquired Tenet partner Neil (Robert Pattison) track the inverted bullets to the nasty, if not sociopathic, Russian oligarch Sator (Kenneth Branagh) who has connections to the hostile future. Through a subplot about art forgery, Protagonist approaches Sator via his wife Kat (Elizabeth Debicki) who hates her husband. Protagonist and Neil discover that Sator can invert people and things with a machine called a Turnstile. As the two probe further, we see scenes with simultaneous forward and backward action, such as a well-choreographed fight between Protagonist future and Protagonist past.

Finally, the scope of Sator's evil becomes clear. He has terminal cancer and plans to destroy the world the moment he dies by triggering the Algorithm, nine devices that together will invert half the Earth. These are hidden back in time and at different locations. Protagonist and his Tenet allies, helped by Kat, cleverly use forward and backward time to disarm the Algorithm and save the world.

Besides other turns in the story, more than can fit here, there are two important blips of science exposition. In a breather from the action, Protagonist and Neil wonder about the "grandfather paradox;" if you travel to the past and kill your grandfather, have you killed yourself too? And in a discussion about reverse chronology, Neil casually throws in that it's like "Feynman's and Wheeler's notion that a positron is an electron moving backwards in time." This is an authoritative claim because, he modestly adds, "I have a Master's in physics."

Backward electrons, entropy, and time travel—how much sense do these ideas make in this film? Most stories that manipulate time come up against temporal paradoxes and what they say about the ability to control our personal fates. From the purely physical viewpoint, cause followed by effect is utterly embedded in science. The fact that travel into the past would violate this chain is taken to mean that traveling backward in time is impossible. If human thought and consciousness are determined solely by physical processes, we cannot do anything in the past either, except remember it. After centuries of philosophical puzzlement over these issues, no science fiction story is going to resolve them, so most do what *Tenet* does: mention them, then ignore them, and return to the action.

Entropy, however, does carry weight in the film because it is linked to the flow of time through the Second Law of Thermodynamics: For any system—say an auto engine—the entropy measures how much of the system's energy is lost to friction-like processes, which turn energy into heat that can never be recovered. The loss grows as the system functions, and so the direction of the increase in entropy defines the way to the future. Entropy has been given the poetic name "the arrow of time" because of this one-way property of systems.

That said, this does not apply to a single particle alone. A video of a speck of dust in motion would not show a qualitative change when run forward or backward, and so would not differentiate past from future. This does not mean that Neil's line about a backward electron is wrong by itself. Richard Feynman, whose mentor was the eminent physicist John Wheeler, did invent a method of calculating how elementary particles would interact if you let an electron go into the past. But this is math that works on paper, not in reality, and Feynman's idea for a single particle, not a whole system, wouldn't qualify as an arrow of time anyway.

If the increasing entropy of a system such as a bullet points to the future, would decreasing its entropy take it to the past? Within the great system of the universe whose entropy is increasing, there are local systems such as your own body where the entropy decreases (this increases the surrounding entropy, so it doesn't violate the Second Law for the whole universe). We don't think of this locally increased entropy as traveling backward in time, however. Entropy may follow the flow of time, but we have hardly begun to know the nature of time itself and whether what drives it is just this thermodynamic principle. Still, *Tenet* gets credit for using entropy as a dramatic marker of time. This is a break from other science fiction stories that use entropy metaphorically to represent a decaying universe, running down to the "heat death" of everything.

Tenet struck many as long on running time and confusion and short on making emotional connections with its viewers. Yet, as Richard Feynman once brilliantly expressed, humanity has its own built-in time arrow that carries feeling: "We remember the past, we don't remember the future. We have a different kind of awareness about what might happen than we have about what most likely has happened." [3] Now there's a topic truly worthy of a humanistic science fiction film about time, entropy, and people.

References

1. Garry Maddox, "The biggest film I've done": Christopher Nolan on the secret world of *Tenet*, *The Sydney Morning Herald*, Aug. 22, 2020. https://www.smh.com.au/culture/movies/the-biggest-film-i-ve-done-christopher-nolan-on-the-secret-world-of-tenet-20200810-p55kd7.html.

2. Nate Jones, A Beat-by-Beat Explanation of What Happens in *Tenet*, *New York*, Sept. 4, 2020. https://www.vulture.com/2020/09/tenet-explained-whats-going-on-in-the-plot-of-this-movie.html.

3. Sidney Perkowitz, Time Examined and Time Experienced, *Physics World*, July 2018. https://physicsworld.com/a/time-examined-and-time%e2%80%afexperienced/.

Books about science for general readers by Sidney Perkowitz

Empire of Light: A History of Discovery in Science and Art (Joseph Henry Press, Washington, DC, 1998, illustrated edition). Hardcopy, paperback and foreign editions, 1996–2005.

Universal Foam: From Cappuccino to the Cosmos (Walker, New York, 2000). Paperback and foreign editions, 2001–2008.

Digital People: From Bionic Humans to Androids (JHP/National Academies Press, Washington, DC, 2004; paperback, 2004). Foreign edition, 2011.

Hollywood Science: Movies, Science, and the End of the World (Columbia University Press, New York, 2007). Paperback, e-book and foreign editions, 2008–2010.

Slow Light: Invisibility, Teleportation, and Other Mysteries of Light (Imperial College Press, London, 2011). Foreign edition, 2014.

Hollywood Chemistry: When Science Met Entertainment, D. Nelson, K. Grazier, J. Paglia, and S. Perkowitz, eds. (ACS Books/Oxford University Press, Washington, DC, 2013; 2014).

Universal Foam 2.0: From Cappuccino to the Cosmos (Kindle e-book, 2015).

Frankenstein: How a Monster Became an Icon, The Science and Enduring Allure of Mary Shelley's Creation, Sidney Perkowitz and Eddy Von Mueller, eds. (Pegasus Books, 2018).

Physics: A Very Short Introduction (Oxford University Press, Oxford, 2019).

Real Scientists Don't Wear Ties: When Science Meets Culture (Jenny Stanford Publishing, Singapore, 2019, c.2020).

Republishing credits

I'm grateful to the following publications and organizations that granted permission for me to republish my works originally published by them, listed below. I am the sole author of each piece.

Sidney Perkowitz

Aeon

Flash!, *Aeon*, May 15, 2019.

Alumni Association, Massachusetts Institute of Technology

Small Wonders, *MIT Technology Review*, July 1993, 70–71.

Astronomical Society of the Pacific

Galileo Through a Lens: Views of His Life and Work on Stage and Screen, in *The Inspiration of Astronomical Phenomena VI*, October 18–23, 2009, Venice, Italy. E. M. Corsini, ed. (San Francisco, CA: Astro. Soc. of the Pacific, 2011), Vol. 441, 85–88.

Creative Loafing

Light in the Woods: Bruce Munro at the Atlanta Botanical Garden, *Creative Loafing*, July 16–22, 2015, 15.

Discover

Codex Futurius: When Houses Grow on Trees, *Discover Science Not Fiction*, Sept. 3, 2009

Emory Medicine

The Better to See You With, *Emory Medicine*, Spring 2019, 23–29.

Insight Editions

Intelligent Machines: It's More Than Just Intelligence, in *James Cameron's Story of Science Fiction,* R. Frakes, ed. (Insight Editions, San Rafael, CA, 2018), 198–204.

Institute of Physics

The Art of Falling Fluid, *Physics World*, April 2014, 52.

The Most Popular Physics Meme Ever, *Physics World*, May 2015, 52.

Paint it Nanoblack, *Physics World*, August 2016, 48.

Physics and Art in 2½ D, *Physics World*, Nov. 2018, 56.

The Physics of Blood Spatter, *Physics World*, Oct. 2019, 43–46.

Getting the Film Physics Right, *Physics World*, Feb. 2020, 56.

From the Lab to the Courtroom, *Physics World*, June 2020, 5.

Interalia

Altered States: 2D Digital Displays Become 3D Reality, *Interalia*, Oct. 2019.

JSTOR Daily

Sociophysics and Econophysics, the Future of Social Science? *JSTOR Daily*, Sept. 26, 2018.

The Quantum Random Number Generator, *JSTOR Daily*, May 22, 2019.

Can Zapping Your Brain Really Make You Smarter, *JSTOR Daily*, Nov. 27, 2019.

How to See the Invisible Universe, *JSTOR Daily*, April 9, 2020.

Labocine Spotlights

Confronting the Wall, *Labocine Spotlights*, March 1, 2018.

Sunlight, Life and Time *Labocine Spotlights*, Nov. 30, 2018.

The Poetry and Prose of Math, Part 1: Poetry, *Labocine Spotlights*, Aug. 25, 2019.

The Poetry and Prose of Math, Part 2: Prose, *Labocine Spotlights*, Sept. 2, 2019.

We and the Earth Breathe Together, *Labocine Spotlights*, Sept. 30, 2019.

Sunday is Maroon: Synesthesia on Screen, *Labocine Spotlights*, Sept. 28, 2020.

Facing Up to Facial Recognition, *Labocine Spotlights*, Sept. 28, 2020.

Science Cinema Online: the 13th Annual Imagine Science Film Festival, *Labocine Spotlights*, Dec., 9, 2020.

Los Angeles Review of Books

A Short Take on Mathematics, *Los Angeles Review of Books*, July 30, 2015.

Bad Blood, Worse Ethics, *Los Angeles Review of Books*, Sept. 7, 2018.

National Academy of Sciences

Aliens: Love Them, Hate Them, or Relate to Them?, *The Science and Entertainment Exchange,* August 2009.

The Chemical Formula: Successfully Combining Chemistry, Science, and the Media, *The Science and Entertainment Exchange*, April 2011.

Representing Robots: Theater First, Film Later, *The Science and Entertainment Exchange*, Aug. 2, 2012.

Nautilus

Most Tech Today Would be Frivolous to Ancient Scientists, *Nautilus Blog: Facts so Romantic,* April 19, 2019.

If Only 19th-Century America Had Listened to a Woman Scientist, *Nautilus*, Nov. 28, 2019.

Only Disconnect!, *Nautilus*, March 26, 2020.

The Power of Crossed Brain Wires, *Nautilus*, June 17, 2020.

The Bias in the Machine, *Nautilus*, Aug. 19, 2020.

A Supermassive Lens on the Constants of Nature, *Nautilus*, Nov. 25, 2020.

Pegasus Books

Frankenstein and Synthetic Life; Fiction, Science and Ethics, in *Frankenstein: How a Monster Became an Icon, the Science and Enduring Allure of Mary Shelley's Creation,* Sidney Perkowitz and Eddy Von Mueller, eds. (Pegasus Books, NY, 2018), 181–204.

Physics Perspective

The Shadow of Enlightenment, *Physics Perspective* **12**, 2010, 234–236.

Sloan Science and Film/Museum of the Moving Image

From Terminator to Black Mirror: Algorithmic Warfare's Perils, *Sloan Science and Film*, Aug. 25, 2020.

Panic in the Streets: Filming the Pandemic, *Sloan Science and Film*, June 23, 2020.

Entropy and the End of the World in *Tenet*, *Sloan Science and Film*, Jan. 27, 2021.

The New York Academy of Sciences

Heat Wave, *The Sciences,* March/April 1992, 30–37.

Mood Indigo, *The Sciences*, March/April 1993, 26–32.

The Washington Post

The War Science Waged, *The Washington Post*, March 3, 1991, c2.

Romancing the Quantum, *The Washington Post*, Oct. 6, 1991, c3.

UNAM (National Autonomous University of Mexico)

Atrapados en Marte (Trapped on Mars), *¿Cómo ves?*, **209**, UNAM, Cd. de México, 2016.

El infinito en el cine (Infinity on Screen), *¿Cómo ves?,* **213**, UNAM, Cd. de México, 2016.

La ciencia avanza y la ciencia ficción le sigue el paso (Science Advances and Science Fiction Keeps Up), *¿Cómo ves?*, **232**, UNAM, Cd. de México, 2018.

Index